Markets of One

The Harvard Business Review Book Series

Designing and Managing Your Career, Edited by Harry Levinson

Ethics in Practice, Edited with an Introduction by Kenneth R. Andrews

Managing Projects and Programs, With a Preface by Norman R. Augustine

Manage People, Not Personnel, With a Preface by Victor H. Vroom

Revolution in Real Time, With a Preface by William G. McGowan

Strategy, Edited with an Introduction by Cynthia A. Montgomery and Michael E. Porter

Leaders on Leadership, Edited with a Preface by Warren Bennis

Seeking Customers, Edited with an Introduction by Benson P. Shapiro and John J. Sviokla

Keeping Customers, Edited with an Introduction by John J. Sviokla and Benson P. Shapiro

The Learning Imperative, Edited with an Introduction by Robert Howard

The Articulate Executive, With a Preface by Fernando Bartolomé

Differences That Work, Edited with an Introduction by Mary C. Gentile

Reach for the Top, Edited with an Introduction by Nancy A. Nichols

Global Strategies, With a Preface by Percy Barnevik

Command Performance, With a Preface by John E. Martin

Manufacturing Renaissance, Edited with an Introduction by Gary P. Pisano and Robert H. Hayes

The Product Development Challenge, Edited with an Introduction by Kim B. Clark and Steven C. Wheelwright

The Evolving Global Economy, Edited with a Preface by Kenichi Ohmae

Managerial Excellence: McKinsey Award Winners from the *Harvard Business Review*, 1980–1994, Foreword by Rajat Gupta, Preface by Nan Stone

Fast Forward, Edited with an Introduction and Epilogue by James Champy and Nitin Nohria

First Person, Edited with an Introduction by Thomas Teal

The Quest for Loyalty, Edited with an Introduction by Frederick F. Reichheld, Foreword by Scott D. Cook

Seeing Differently, Edited with an Introduction by John Seely Brown

Rosabeth Moss Kanter on the Frontiers of Management, by Rosabeth Moss Kanter

Ultimate Rewards, Edited with an Introduction by Stephen Kerr

Peter Drucker on the Profession of Management, by Peter F. Drucker

On Competition, by Michael E. Porter

The Work of Teams, Edited with an Introduction by Jon R. Katzenbach

Delivering Results, Edited with an Introduction by Dave Ulrich

John P. Kotter on What Leaders Really Do, by John P. Kotter

Creating Value in the Network Economy, Edited with an Introduction by Don Tapscott

Managing in the New Economy, Edited with an Introduction by Joan Magretta

World View, Edited with an Introduction by Jeffrey E. Garten

Markets of One, Edited with an Introduction by James H. Gilmore and B. Joseph Pine II

Markets of One

Creating Customer-Unique Value through Mass Customization

Edited with an Introduction by
**James H. Gilmore and
B. Joseph Pine II**

A Harvard Business Review Book

The *Harvard Business Review* articles in this collection are available as
individual reprints. Discounts apply to quantity purchases. For information
and ordering contact Customer Service, Harvard Business School Publishing,
Boston, MA 02163. Telephone: (617) 783-7500 or (800) 988-0886, 8 A.M.
to 6 P.M. Eastern Time, Monday through Friday. Fax: (617) 783-7555, 24
hours a day.

Library of Congress Cataloging-in-Publication Data

Markets of one : creating customer-unique value through mass
    customization / edited with an introduction by James H. Gilmore and
    B. Joseph Pine II.
        p. cm. — (A Harvard business review book)
    Includes index.
    ISBN 1-57851-238-7 (alk. paper)
    1. Flexible manufacturing systems. 2. Product management.
3. Relationship marketing. I. Gilmore, James H., 1959– . II. Pine, B.
Joseph. III. Series: Harvard business review book series.
TS155.65.M37 2000   99-32967
670.42'7—dc21   CIP

The paper used in this publication meets the requirements of the American
National Standard for Permanence of Paper for Publications and Documents
in Libraries and Archives Z39.48–1992.

Contents

Introduction:
Customization That Counts

James H. Gilmore and B. Joseph Pine II

Roll over, Henry Ford. Today, you *can* have any color you want, as long as it's the one you want. Technologies enabling mass customization now permeate a vast number of manufacturing and service industries, from automobiles, blue jeans, and mattresses to grocery shopping, insurance, and personal investing. Mass customization represents no mere business fad, but an ongoing and inexorable shift in the very structure of American economic activity—just as mass production was in its day. As Robert T. McTeer, Jr., President and Chief Executive Officer of the Federal Reserve Bank of Dallas, highlighted in his organization's 1998 annual report, *The Right Stuff: America's Move to Mass Customization*:

> Things used to be made to order and made to fit. But they were labor-intensive and expensive. Mass production came along and made things more affordable, but at a cost—the cost of sameness, the cost of one-size-fits-all.
>
> Technology is beginning to let us have it both ways. Increasingly, we're getting more personalization at mass-production prices. We're moving toward mass customization.[1]

The Texan should know—his state is the home of the Henry Ford of mass customization, one Michael Dell, whose company's direct-to-the-customer, build-to-order, virtual-distribution capabilities are the envy of every manufacturer in the PC industry—and nearly every other industry as well. (The *Wall Street Journal* reports that "to dell" has become a verb in today's business parlance.[2]) Rather than mass-produce computers to place in inventory for potential market demand, Dell Computer mass-customizes—efficiently producing output only in response to actual orders—and as a result, enjoys a considerable cost advantage relative to its competition. Dell's assembly process uses standardized *modules*—the keystone of mass customization—and combines them in different ways to make PCs configured to the specific needs of individual buyers. And it is this mass-customizing capability that allows Dell to "go direct" to its customers in such an impressive way, financing its growth with no working capital since its customers, on average, pay their bills more than a week before Dell has to pay its suppliers.

The very existence of these *personal* computers bears witness to the sea change from mass production to mass customization. When the term "personal computer" was first coined (attributed to Xerox PARC's Alan Kay, now at Walt Disney Imagineering), computers were the size of automobiles, or bigger. They were giant machines processing data in the foreboding basements of only the largest of institutions. Today, we carry them in the smallest of briefcases. And we increasingly take this anytime/anyplace computing for granted.

Indeed, what the automobile did in the twentieth century to change the landscape of American commerce and culture—indeed that of the entire industrialized world—the computer is doing as we enter the twenty-first century. It is changing how both businesses and consumers think about the very fabric of our economy. Ford's assembly line, and the mass-production mind-set it fostered, led to a conception of customers aggregated into mass markets. Providers of goods and services came to design, produce, and deliver standardized products at low enough prices that nearly everyone could afford to buy them. In product category after product category, as new mass-produced items rolled off the lines, most consumers gladly sacrificed what they wanted exactly in order to simply obtain one.

Fast forward. Today, individual customers need no longer rely on standardized products to meet their unique needs. To buy a computer, many businesses and consumers now call Dell's toll-free number or, increasingly, go to its Web site and order a PC made to their exact specifications. Or they call rival Gateway for similarly mass-customized PCs. And as people come to enjoy customized offerings in some facets of their lives, at prices they are willing to pay, they begin to expect them in other facets as well.

Indeed, mass customization will be as important to business in the twenty-first century as mass production was in the twentieth. Even the automobile industry is starting to get on the bandwagon: In Europe, you can order a Smart Car from Mercedes with interchangeable color panels installed at the dealership, and in the U.S. Saturn will build a car just for you without taking three months. Tired of jeans that don't fit right, or of buying the same brand or size as last time only to find that they don't fit at all? (It may be that you did *not* gain weight, for inconsistencies are inherent to cutting scores of sheets of denim at one time, as they do when mass-producing jeans.) Go buy a pair of mass-customized jeans from Levi's, or—with more options but somewhat higher price—from the Interactive Custom Clothes Company at www.ic3d.com. Married to someone who prefers a different level of firmness in a mattress? Separate beds are not the answer (at least not for most couples). Instead, buy a Select Comfort mattress that enables each of you to independently control the firmness on your side of the bed. To avoid weekly shopping trips for groceries, become a Peapod or Streamline customer, create your own on-line grocery store, and order home delivery. Just undergo a significant life event, such as selling a house or having a baby? If you're fortunate enough to be a member of USAA, watch it customize a set of offers to your particular situation. Bombarded with statements from all your different mutual fund investments? Open a Charles Schwab account and get them all listed on a single report (and feel free to add or drop other brands of funds to your portfolio—with still one statement).

While most of the prominent examples of mass-customized offerings may be in the consumer arena, business-to-business companies have embraced mass customization just as much, if not more. Just a sampling of such companies includes Aramark in hospital services, Wildfire in telecommunication services, Cisco in telecom equipment, Ross Controls in pneumatic valves, ChemStation in industrial soap, and Lutron Electronics in lighting controls. And many companies, such as Lutron and Dell, encompass both business and consumer markets.

These companies tailor their products to meet the unique needs of individual customers in such a way that nearly all can find exactly what they want at a reasonable price (often about the same as for mass-produced products). The companies have achieved this only after embracing a new mind-set, one of creating *customer-unique value*

through mass customization. Their new offerings acknowledge an ax-
iom all too easily ignored in the era of mass production: Every customer is unique. Period. And with every customized sale, a slice of the
mass-market paradigm falls away. Slowly but surely, the way we've
come to think about markets over the past century fades from view. In
its stead, a new conception of markets emerges, one that recognizes
the obvious: that every individual customer is a *market of one*.

This shift in how we think about markets is precisely what Stan Davis discussed when he first coined the term *mass customizing* in his seminal 1987 book, *Future Perfect*.[3] As Davis explained, before the Industrial Revolution, the dominant market paradigm was the local market,
where producers of goods and providers of services generally operated
within limited geographical boundaries. The common conception of a
market was of a physical place where buyers and sellers came together
at a specific time and place to exchange money for product.

The notion of mass markets only emerged after the Industrial Revolution, when mass producers began to standardize goods and services.
To embellish Say's Law, standardized supply created its own homogeneous demand. This worked very well in most cases, until the forces of
competition brought about new ways of thinking about markets and
how to serve them. In the automobile industry, for example, Alfred
Sloan outcompeted Henry Ford by segmenting the mass market based
on sociodemographic factors and putting in place a distinct car company—Cadillac at the high end, then Buick, Oldsmobile, and so on—
to achieve nearly the same economies of scale Ford had for the whole
market. In industry after industry, this market segmentation led to the
drawing of finer and finer distinctions between groups of customers,
until the market niche was born.

What Davis foresaw was the next natural progression in markets,
from the niche to what he called "mass-customized markets," or markets of one, where each individual customer, whether a consumer or a
business, would be his own market. We are now at such a time. We
now see anew that each customer is unique, and no longer has to sub-
sume that uniqueness into a homogeneous market in order to get a
decent price for a product.

This book, then, chronicles the evolution of business competition as
seen in the pages of the *Harvard Business Review*—from mass markets
to markets of one, from creating standardized value through mass production to creating customer-unique value through mass customization. The articles we've chosen span the last dozen years, but clearly

point to how businesses will compete in the coming decades. These articles are foundational to any understanding of how to embrace mass customization.[4] While we here review only the most salient points of each—as well as add a few never-before-published points of our own—any reader wishing to compete in this new century should delve fully into the details of each, beginning with a piece by that most prolific of contributors, Peter Drucker.

The Demise of Mass Markets

Drucker embodies the spirit of that old E.F. Hutton brokerage commercial: When he talks, people listen. And in listening to his voice, new business practices emerge. His every contribution to the *Harvard Business Review* contains great nuggets of wisdom, forever altering how we think and manage. "The Emerging Theory of Manufacturing" would, at first glance, appear to be one of Drucker's more narrowly focused HBR articles, and indeed the piece was his 1990 specification of the future factory, circa 1999. Looking back over the past ten years, however, one sees the breadth of the article's scope, for the new reality of which Drucker spoke at the time he wrote this article now reigns in manufacturing and nonmanufacturing industries alike. In a sense, his emerging theory delivered a death sentence to the mass production mind-set that had for so long influenced all business thinking.

As Drucker stated, "manufacturing people tend to think like Henry Ford: you can have either standardization at low cost or flexibility at high cost, but not both." Such reasoning, of course, was not limited to those who worked in manufacturing; many service industry managers shared the same point of view. Service businesses also sought to stabilize nearly everything they did in order to efficiently generate uniform output for the mass market; or they produced higher-priced variations for market niches. Since at the time both goods and services companies embraced this manufacturing mind-set, Drucker's article was aptly named.

The four "manufacturing" concepts that Drucker outlined in 1990 have indeed transformed the way nearly every business competes. Statistical quality control and other continuous improvement tools are now practiced on both the factory floor and the service corridor. Quality is a given, providing—as Drucker called it—a new "social organization" for contemporary business. *Either/or* trade-offs are no longer socially acceptable, in the boardroom or in the bowels of an enterprise. Today, businesses seek *both/and* solutions at every turn; the mutually exclusive has become the mutually assured. We have W. Edwards Deming, Joseph Juran, Phil Crosby, et al. to thank, for the enduring legacy of the quality movement is the now commonly understood realization that you can have both low costs and high quality.

Drucker also anticipated a corresponding change that occurred in the 90s in how businesses made decisions: "The beans will be counted differently," he said. And he was right again. Cost accounting has been turned on its head. Before, inventory was free and time was plentiful. Now, time is scarce and inventory must be avoided. Why the change? Simply put, businesses realized that the production factory—and similarly, any service operation—could not be accounted for or managed independently from the rest of the enterprise. And so, manufacturing has been integrated into business strategy, along with other business processes and the technologies that enable new forms of work to be performed. Innovation is now the simultaneous rethinking of business strategy and business operations. Companies must weigh not only the cost of doing something, but also of *not* doing that same thing. For if you don't do it, someone else will.

The third concept that Drucker identified was that of the "flotilla" as a means of organizing work—that is, having a collection of capabilities that are constantly reorganizing. Indeed, such modular capabilities, as nearly every article in this collection attests, have become the new imperative for both the product and process architectures of any enterprise. Business change *built atop* mass-production processes simply preserves the old either/or regime, and in the end just adds costs. On the other hand, the module is the basic building block for developing both/and capabilities. A portfolio of modules—combined with a system to dynamically link selected modules together on demand—provides the means to mass-customize. Flexibility is *built into* the architecture itself. As a result, the delivery of goods, as Drucker points out, "no longer functions as a step-by-step process that begins at the receiving dock and ends when finished goods move into the shipping room. Instead, the plant must be *designed from the end backwards* and managed as an integrated flow" (our emphasis). Static linkages give way to dynamic flows. Supply chains are replaced with demand chains, with the focus placed where it belongs—on the end customer.

Drucker's fourth and final concept emanated from this new focus. No longer product-focused or centrally controlled, successful businesses began to sense and respond more quickly to changing market needs, lest competitors beat them to the task. In such a world, companies no longer viewed themselves as merely making or doing things, but as instruments to create value for their customers. This "systems approach," as Drucker called it, had profound implications for the very function of the corporate organization. Indeed, as Drucker correctly predicted, "the factory of 1999 will be an information network," with information about individual customers driving operations.

Regis McKenna extends this view from operations to marketing. In "Marketing in an Age of Diversity," he argues that this new economic order means "abandoning old-style market-share thinking and instead tying the uniqueness of any product to the unique needs of the customer." With the erosion of mass-produced offerings came the demise of mass-market thinking, as the variety of goods and services available to individual customers exploded throughout the late '80s and the '90s. To demonstrate this phenomenon, McKenna cites such statistics as the growth of companies in the semiconductor industry and the number of grocery items stocked in an average supermarket. You can validate this happening within your own industry, not just by counting the total number of products offered, but by calculating a more revealing metric: what we call the *variety index*. Simply take the total number of products offered in your industry during a given year (or that of just your own company, if only those data are available), and then divide that number by the market share of the best-selling product. Do this, say, for each year in the '90s, and plot your industry's variety index over time. In every case we've seen, this variety index—representing the proliferation of one's offerings—grows exponentially, for the numerator increases as the denominator decreases with each passing year. "Gone is the convenient fiction of a single, homogeneous market," McKenna declares.

Arguing which came first—the migration away from mass-production practices on the supply side or the fragmentation of mass markets on the demand side—is a pointless chicken-or-egg debate. There couldn't be one without the other. As companies embraced quality principles, focused on creating value (not just products), and challenged traditional assumptions about how to create that value, they introduced unprecedented levels of variety into the marketplace. And as individual customers began to purchase new offerings that more closely approximated exactly what they wanted, their expectations changed. If one obtained exactly what one wanted in one product category, then one demanded similar treatment in other categories. And this demand led providers of goods and services to introduce additional offerings, fostering further market fragmentation.

What matters is where we've ended up: with a set of conditions that McKenna amazingly and accurately portrayed (in 1988) at the dawn of this change. We now must think of customers as individuals (in fact, the phrase "individual customer" is a redundant one, needed only to remind us to escape the mass-market mind-set of yesteryear). Customers cannot be conveniently categorized into aggregate market groupings; one must interact one-to-one with customers to ascertain their needs. Companies cannot easily or accurately predict individual tastes and preferences; traditional surveys and focus groups no longer suffice and must be supplemented or replaced altogether with direct observations or dialogue-based research. More and more ads reach fewer and fewer households; more and more households resemble fewer and fewer other households. Market share is passé, and share-of-customer (and the derivative share-of-wallet, share-of-household, and share-of-anything-individual) metrics rule the day. Quite simply, executives can no longer sit back and ask, "Is there a market for X or Y?" Rather, markets must be proactively created by customizing one's capabilities to match the increasingly heterogeneous needs of customers.

Indeed, the demise of mass markets requires companies to drastically redefine how they think about their "product." Customizing a good automatically turns it into a service, and customizing a service automatically turns it into an *experience*—a memorable event that engages a customer in an inherently personal way. As goods and services increasingly become commoditized, customers place more value on the experience they receive from companies.[5] Mass customization is clearly a route to staging that experience, one that two giant thinkers, Drucker and McKenna, help us see by replacing the familiar spectacles of mass production and mass marketing with lenses that look not at the commonality across all customers but at the uniqueness that exists in each and every individual.

Efficiently Serving Customers Uniquely

In their "Managing in an Age of Modularity," Carliss Baldwin and Kim Clark correctly state that modularity increases the rate of innovation in an enterprise. They go on to outline many of the benefits of modular product and process structures: enabling greater complexity to be handled efficiently, intensifying vendor competitiveness, and accelerating a company's ability to move into and out of new businesses, among others. They also ask this very insightful question: "If modularity brings so many advantages, why aren't all products (and processes) fully modular?" Their answer: Modular capabilities are much more difficult to design than integrated products. But the advantages are great, for modularity enables companies to efficiently serve customers uniquely.

Conversely, it is also true that in order to focus on unique customer needs, modular capabilities are a necessity. However, establishing such capabilities requires some down-and-dirty dealing with various operational details of one's business. As Baldwin and Clark point out, "The designers of modular systems must know a great deal about the inner workings of the overall product and process in order to develop the visible rules necessary to make the modules function as a whole." To add to the challenge, those individuals and groups within an organization that have detailed knowledge of how things work are frequently the keepers of the current paradigm and are too often unwilling or reluctant to embrace change. So designers of modular capabilities must not only redesign work, they must also reorganize their company's resources (or organize new businesses within companies) to achieve customization prowess. The authors know this well, and acknowledge the need to articulate a compelling vision and to "delineate and communicate a detailed operating framework" that brings the vision into practical terms for those affecting and affected by the change.

Baldwin and Clark provide a valuable first step in developing such a framework by urging companies to distinguish *hidden design parameters* (those that do not influence the design beyond the individual module) from *visible design rules* (those that influence subsequent design decisions). They also outline three visible design rule categories: *architecture* (or portfolio of modules), "which specifies what modules will be part of the system and what their functions will be"; *interfaces* (or the linkage system) "that describe in detail how the modules will interact"; and *standards* "for testing a module's conformity to design rules." These categories provide a valuable framework. Furthermore, we believe that specifying the nature of each module goes a long way toward describing the interfaces, and vice versa. That is, the more detailed the specification of the modules, the more obvious the interface issues; and the more the interfaces are clearly understood, the more insight one has about the functional requirements of individual modules.

When pursuing modular architecture and interfaces, we suggest you address the following characteristics (or descriptors) of the modules, amongst other possibilities:

- Connectivity—defining all the modules that *can* precede, follow, or operate in parallel with each module
- Dependency—defining all the modules that *must* precede, follow, or operate in parallel with each module
- Technical linkage—defining how each module can be linked to preceding, following, and parallel modules (whether via human handling, expert system, configurator, communication technology, or other methods)
- Performer—defining which individuals can perform activities in each module
- Place—defining where activities in each module can be performed (for both physical and virtual space)
- Communication mode—defining how the status of the module is communicated during the linkage process (proactively, exception-based, look-up, etc.)
- Ownership—defining who has responsibility for managing the activities within each module

Note that a module is a collection of components (for goods) or activities (for services) that share a common set of descriptors and are assembled or performed as a single unit of material or work. Issues of *granularity*—think of Lego building blocks and how they come in differently sized pieces—address the degree to which individual product parts or service activities operate as distinct modules. It is not necessary that all modules be highly granular—that is, that the individual parts or activities within them be modules unto themselves. Necessity and efficiency often require less granularity to achieve greater modularity. The key is to establish a *dynamically* linked collection of modules that create customer-unique value in ways not possible through the statically linked processes of mass production (and continuous improvement for that matter).

Goods and services businesses differ in the challenges faced in establishing such modular capabilities (and yet additional challenges exist when staging experiences, when the architecture and interfaces of performance are best understood in terms of street theatre).[6] Because goods are tangible, modules and their descriptors are easier to conceive and describe—one can see and touch the components. But the physicality of modules can pose linkage difficulties, often requiring costly retooling of manufacturing processes (another reason we don't see more modular goods). Services, however, are inherently customizable. Because they're intangible, service modules are much easier to dynamically link in different combinations and sequences for different customer needs. But with services, conceptualizing the nature of the modules often becomes much more difficult because designers and work teams have no tangible components or prototypes to physically examine.

In any case, why bother migrating to mass-customizing capabilities? Because, in simple terms, the potential to differentiate one's offerings is too great to ignore. Consider the advice of Stan Davis in a letter to the editor in HBR (March–April 1994, p. 178): "It is worth asking which elements in the product-market mix do you want to mass-customize. And where along the value chain do you want to mass-customize—in design, manufacture, sales, service? Be selective: a good rule of thumb is to mass-customize as much as necessary and as little as possible."

Indeed, companies should customize only where it counts. Too much customization—like too much variety—can overwhelm customers with too much choice. It is possible to inundate customers with features and benefits beyond that which customers find meaningful. "Do You Want to Keep Your Customers Forever?" cowritten by Joe Pine, Don Peppers, and Martha Rogers, offers an invaluable perspective on selecting just what to customize. In particular, one small paragraph—almost hidden amidst all the other practical advice—weighs in on this matter and offers a new performance metric, *customer sacrifice*, defining it as "the gap between what each customer truly wants and needs and what the company can supply." This gap is the place to start customizing, the authors say—where customers experience the greatest sacrifice, or what the Federal Reserve's McTeer called "the cost of one-size-fits-all."

The purpose of customizing must be to decrease customer sacrifice, and to decrease it at least enough that customers are wooed away from competing offerings and locked into the customized alternative. The purpose of customizing then is not merely to increase customer satisfaction—to lessen the gap between what each customer expects and what he perceives to get from the company. That traditional measure is no longer enough, since expectations are conditioned by years, indeed decades, of settling for something less (and sometimes something more) than what each customer wants exactly. Instead, companies must identify dimensions of sacrifice that, were they to be reduced or eliminated, would create significant value for each customer.

The best way to begin doing that, Pine, Peppers, and Rogers argue, is to form *learning relationships* with customers, for "every interaction with a customer is an opportunity to learn" about how individuals sacrifice. This represents a very different form of interaction with customers than that associated with mass-production and mass-marketing practices, which in response to fragmenting markets simply push more and more options into distribution and more and more messages into more and more media. Customers are left to fend for themselves. With learning relationships, the company seeks to offer *only and exactly* what each individual customer desires, and to do so via direct dialogue with the customers themselves.

But, as Stan Davis pointed out in his letter, one cannot do so in every case. Indeed, mass customization does not mean being everything to everyone. Rather, it means identifying and focusing on a few areas of sacrifice (or perhaps only one) and developing modular capabilities that address those unfulfilled needs. Leading customizers do exactly that. For example, Levi's found that most people sacrificed whenever they bought jeans off the rack, relying on one or two numbers (size for women, waist/inseam for men) to determine every dimension of the pair of jeans. So Levi's determined those dimensions along which consumers sacrificed the most—waist, hips, inseam, and rise (how high the jeans are worn on the waist)—and began matching consumers to modular patterns to minimize sacrifice. In order to compete, another clothing retailer, the entrepreneurial Interactive Custom Clothes Company (IC3D), more than tripled the number of dimensions along which it would eliminate sacrifice (such as circumference around the knee), reduced the unit of measure from the one-half inch Levi's used to any specifiable measure, and developed a program to create a unique pattern for every individual customer (which then enabled it to do things like make pants legs different lengths on the same pair of jeans). Unable to compete with Levi's on points of contact with consumers, the start-up went onto the World Wide Web to enable customers to access its mass-customization capabilities.

As IC3D demonstrates, the possibilities of which elements in your offering—and where in your value chain—to customize are vast. There are often hundreds if not thousands of dimensions to consider, especially if you apply this thinking to the world of e-commerce. As Pine, Peppers, and Rogers state, "anything that can be digitized can be customized." We believe electronic commerce should be customized commerce—at least if one wants to use the Internet as a means to add value to one's offerings. Simply replicating on-line one's atom-based offerings commoditizes the products, whereas a mass customizer will surely exploit the nearly instantaneous and friction-free availability of on-line data to find "products for customers, not customers for products." Once your offering—or, as with IC3D, some appropriate representation of your offering—is digitized, the possibilities are virtually endless. So when developing your e-business, think about what sorts of customer sacrifice can only be addressed in the digital world and build customization capabilities around them.

On-line or off, there is much to do if one truly wishes to create customer-unique value. In "Is Your Company Ready for One-to-One Marketing?" Peppers and Rogers, with Bob Dorf, provide a plethora of practical questions to ask oneself before embarking on a customization campaign. One would be wise to closely scrutinize the tools and templates delineated in this article. The authors also outline an approach to justify the investment in new customization capabilities, namely to examine *customer skew*—or how widely the value of individual customers varies within your business. As they state, pursuing customization and one-to-one marketing "is more cost efficient for businesses with a steep skew than with a shallow one. The greater the skew, the more feasible it is to cultivate relationships with the most valuable customers. If the top 2% of your customers generate 50% of your profit, you can protect 50% of your bottom line by fostering learning relationships with just the top few customers."

The logic is sound, and many new initiatives can and should be financed on that basis. But since lifetime values of customers are often calculated based primarily on past purchasing behavior, be sure not to look solely at current customer data. Even when potential changes in future buying behavior (such as increased purchases due to receiving customized treatment) are taken into consideration, the focus should remain on *all* buyers. Attention should also be given to those who are not presently customers. Why are they *not* buying today? Examining that question may identify some dimension of sacrifice that is relatively unimportant to existing customers (or especially to the top few customers that comprise the majority of your past revenue) but more important to many prospective customers.

In fact, we encourage companies to understand *sacrifice skew*—or how widely the sacrifice of individual customers varies across different dimensions of existing offerings. Look for the "common uniqueness" found in *all* customers in your industry (or other industries to which you might migrate). That is, what are those dimensions of your offering in which what customers want *exactly* varies the most? The greater the skew in a particular dimension, the more valuable it would be to customize that one dimension for all customers, and the greater the return on developing new capabilities. If just one feature—among currently available offerings—has generated great sacrifice for everyone, customizing that feature could attract many new customers to your business (and perhaps lower your customer skew!).

The Business of Mass Customization

Creating customer-unique value should not just protect existing businesses, but grow new ones. In "Breaking Compromises, Breakaway Growth," George Stalk and his colleagues David Pecaut and Benjamin Burnett demonstrate that eliminating customer sacrifice offers a means for companies to generate added revenue. For it is clear that innovation focused on sacrifice reduction (we might ask: Is there any other genre of innovation?) creates new demand. That is the business of mass customization. In most, if not all, industries, customers refrain from spending more money—by not buying at all, postponing repurchases, or buying a lower-priced offering—because existing options require them to compromise by accepting something other than what they want exactly.

This compromise, of course, equates to customer sacrifice. Listen to Stalk, Pecaut, and Burnett describe compromise: It "is a concession demanded of consumers" where an industry "imposes its own operating constraints on customers." In other words, it's a kinder and gentler term for the same forfeiture of customers' true desires before the altar of mass production. Despite the softer terminology, Stalk and company take time to detail the harsh reality these compromises "force" customers to "endure." Their observations capture the essence of the beast.

First, customers universally make these sacrificial concessions because they typically have no choice. Second, most of these take-it-or-leave-it predicaments are "hidden" from the customer. Customers have grown so accustomed to established industry practices that they no longer expect *any* alternative—let alone one that's exactly what they want. The sacrifice only "becomes visible when customers have to modify their behavior" to use a good, service, or even experience. And therein lies a great impediment to innovation: The sacrifice upon which innovation is often based has been hidden from the very company providing the existing offerings. That is why innovation, as we're frequently reminded, so often comes from outside one's industry, from one who discovers the sacrifice being made.

That's also why we prefer the stronger language of sacrifice. We've been conditioned to view "compromise" as a good thing; only extremists resist it. Yet extremism in the pursuit of reducing customer sacrifice is no vice. Breaking the long-established patterns of behavior within an industry requires extreme action. There are customers sacrificing out there, and some company ought to break away from the herd mentality of the mass-producing pack and do something about it. And when one does, it often finds customers willing to pay a premium for the sacrifice reduction. The competition stands back asking themselves, why didn't we think of that?

The business of mass customization begins with this separation from mass producers and an exploration of the possibilities for creating customer-unique value. The purpose of such exploration is to exploit a few areas of sacrifice in order to create increased value. But finding those few areas is no easy task. We suggest you begin with a rich exploration of current touchpoints with customers, literally *mapping* the sacrifice. Refrain from assessing "How are we doing?" and instead ask, "What do customers really want?" Avoid thinking in lowly terms of satisfying customers or meeting expectations; rather, push aspirations to the extreme and thereby create a "World Atlas of Sacrifice." Simply naming first the continents of sacrifice and then the countries and finally the cities of sacrifice in your industry and arranging them in relationship to one another can make explicit important problems that companies never find in satisfaction data. Then ask yourself: What sacrifice, if eliminated, would represent the greatest value to customers?

The follow-up question, of course, is then: What might we do to eliminate that sacrifice? The development of such innovative solutions is ultimately what counts. Our own "The Four Faces of Mass Customization" outlines four approaches. Each involves the interplay between the identification of sacrifice and the conceptualization of possible remedies. In this exploring and envisioning process, we take exception to one assertion made by Stalk, Pecaut, and Burnett, namely that trade-offs experienced by customers should not be examined. Their definition of trade-offs—"the legitimate choices customers make among different . . . offerings"—necessitates their position. But what of those *faux trade-offs,* seemingly inescapable but in fact only the by-product of mass-production processes? Companies can use *collaborative customization* to resolve trade-offs that customers make unnecessarily because they've never been offered a customized alternative. Collaborative customizers work directly with individual customers to help them uncover these *either/or* sacrifices in order to identify the precise offering that fulfills those underlying needs, and then provide the customized output.

The other three customization approaches address different forms of sacrifice. Rather than provide a customized offering, with *adaptive customization* one offers a standard, but customizable, product that is designed so that users can easily tailor, modify, or reconfigure it themselves. This approach is appropriate when one "set-up" eliminates the *sort-through* sacrifice needed to find what one wants amid the variety of mass-produced choices. *Transparent customization* provides individual customers with customized goods, services, and experiences, without letting them know explicitly that those offerings have been uniquely configured for them. This approach makes sense when customers wish to avoid any *repeat-again* sacrifice, such as taking the time to describe their unique needs each time they interact with a company. In such situations, customers value customization—they just don't want to be bothered collaborating in order to get it. Finally, with *cosmetic customization,* a standard offering is packaged specially for each customer. This type of customization makes sense to eliminate the *form-of* sacrifice inherent to standardized packaging, marketing materials, product placement, terms and conditions, and other outward representations of the offering.

These four forms of customization emerge from decoupling one's view of product (what it really is) and the representation of product (what one says it is). Customization solutions need not necessarily focus on customizing the product itself; often customizing the information *about* the product can alone create tremendous customer-unique value.

This type of customization is central to Carl Shapiro and Hal Varian's "Versioning: The Smart Way to Sell Information." By examining what they call "information products," those *offerings that consist entirely of representations*, they provide additional insights about serving markets of one.

Remember that information in itself is not a product. As our friend John Perry Barlow likes to say, information wants to be free. Only when the information is packaged in the form of informational *goods*, information *services*, and informing *experiences*, does it yield economic value. Note that each of the points that Shapiro and Varian make about these information products apply to other products as well, and in particular to their representation: First, one product can exist in many versions, especially when packaged in digital form. Second, this "versioning" does not require that a unique product (or version of it) exist for each and every customer. Different customers often select identical versions to fulfill their unique needs.

Shapiro and Varian recommend a very practical approach for developing the dynamic capability required for fulfillment. They suggest that mass customizers first develop the most advanced version possible, and then subtract features that any one customer does not want. This highlights a popular misconception of mass customization. Creating customer-unique value through customization does not always mean *adding* new and different features and benefits. Indeed, *removing* elements from existing offerings can be equally valuable, for customers sacrifice not just in getting too little but in having too much foisted on them—think, for example, of VCRs with so many functions that customers can't figure out how to set the time.

In "Making Mass Customization Work," Joe Pine, Bart Victor, and Andrew Boynton address a number of other misconceptions. First they point out that "mass customization is not simply continuous improvement plus." Offering more and more choices of better and better features does not automatically translate into providing customers with what they want. And thus it follows that "variety in and of itself is not necessarily customization." As we've seen, mass-customized capabilities require a fundamentally different way of operating one's business.

Pine, Victor, and Boyton also provide insight about the pitfalls that many companies encounter in their pursuit of mass customization. Using the lessons learned by Toyota Motor Company—where "product proliferation took on a life of its own"—to launch their analysis, they detail the process flow, information technology, knowledge management, and organizational structures needed to transform work for mass-customization prowess. To achieve such capabilities, companies "must break apart the long-lasting, cross-functional teams and strong relationships built up for continuous improvement to form dynamic networks." Read this installment carefully, for the differences between craft production, mass production, and continuous improvement with mass customization are precisely enumerated. This is especially important for companies employing new technologies (and who isn't today?). As Pine, Victor, and Boynton point out, "many managers still view the promises of advanced technologies through the lens of mass production."

Stephan Haeckel and Richard Nolan offer an alternative lens for viewing technology in "Managing by Wire." Drawing an analogy from aviation, they see the management of technology today as rather like *flying by wire*—where the computer system is used to augment rather than automate the pilot's work. Interestingly, "when pilots fly by wire, they're flying informational representations of airplanes." Likewise, executives of a retail chain manage informational representations of stores, those of a production facility manage informational representations of the plant, and so forth. Information about what one's business is doing *right now* becomes the platform for changing how one does business—on the fly.

The implications, as Haeckel and Nolan note, are profound. Businesses must develop networked IT capabilities and not just isolated IT systems. These capabilities must "be used to create and manage building blocks of business activity that can then be combined into a variety of responses." Responding flexibly to the unique needs of customers must be institutionalized—and not just with daily operations. As Haeckel and Nolan see it, "adopting a manage-by-wire strategy is nothing less than a change in the nature of strategy itself, from a *plan* to produce specific offerings for specific markets to a *structure* for sensing and responding to change faster than competition." That's certainly a challenging assignment, and a worthwhile ambition. And ad hoc leadership won't suffice to steer your company in this way. Someone needs to take the lead, and then lead indeed, if such a business is to offer customization that counts.

Notes

1. Robert D. McTeer, Jr., "A Letter from the President," in *The Right Stuff: America's Move to Mass Customization*, Federal Reserve Bank of Dallas, 1998 Annual Report, p. 1.

2. "Corporate Caveat: Dell or Be Delled," *Wall Street Journal,* May 10, 1999, p. A1.

3. Stanley M. Davis, *Future Perfect* (Reading, MA: Addison-Wesley Publishing Company, Inc., 1987).

4. Interestingly, although its focus lies elsewhere and so is not reprinted here, the very first mention of the term "mass customization" in the pages of HBR was in a piece by John Seely Brown of Xerox, "Reinventing the Corporation," January–February 1991.

5. See B. Joseph Pine II and James H. Gilmore, *The Experience Economy: Work Is Theatre & Every Business a Stage* (Boston: Harvard Business School Press, 1999).

6. Ibid., Chapter 7, and in particular endnote 24 on pp. 221–223, which explains how street theatre equates to mass customization.

PART

I

The Demise of
Mass Markets

1
The Emerging Theory of Manufacturing

Peter F. Drucker

We cannot build it yet. But already we can specify the "postmodern" factory of 1999. Its essence will not be mechanical, though there will be plenty of machines. Its essence will be conceptual—the product of four principles and practices that together constitute a new approach to manufacturing.

Each of these concepts is being developed separately, by different people with different starting points and different agendas. Each concept has its own objectives and its own kinds of impact. Statistical Quality Control is changing the social organization of the factory. The new manufacturing accounting lets us make production decisions as business decisions. The "flotilla," or module, organization of the manufacturing process promises to combine the advantages of standardization and flexibility. Finally, the systems approach embeds the physical process of making things, that is, manufacturing, in the economic process of business, that is, the business of creating value.

As these four concepts develop, they are transforming how we think about manufacturing and how we manage it. Most manufacturing people in the United States now know we need a new theory of manufacturing. We know that patching up old theories has not worked and that further patching will only push us further behind.

Author's note: I wish to acknowledge gratefully the advice and criticism I received on this piece from Bela Gold and Joseph Maciariello, friends and colleagues at the Claremont Graduate School.

Together these concepts give us the foundation for the new theory we so badly need.

The most widely publicized of these concepts, Statistical Quality Control (SQC), is actually not new at all. It rests on statistical theory formulated 70 years ago by Sir Ronald Fisher. Walter Shewhart, a Bell Laboratories physicist, designed the original version of SQC in the 1930s for the zero-defects mass production of complex telephone exchanges and telephone sets. During World War II, W. Edwards Deming and Joseph Juran, both former members of Shewhart's circle, separately developed the versions used today.

The Japanese owe their leadership in manufacturing quality largely to their embrace of Deming's precepts in the 1950s and 1960s. Juran too had great impact in Japan. But U.S. industry ignored their contributions for 40 years and is only now converting to SQC, with companies such as Ford, General Motors, and Xerox among the new disciples. Western Europe also has largely ignored the concept. More important, even SQC's most successful practitioners do not thoroughly understand what it really does. Generally, it is considered a production tool. Actually, its greatest impact is on the factory's social organization.

By now, everyone with an interest in manufacturing knows that SQC is a rigorous, scientific method of identifying the quality and productivity that can be expected from a given production process in its current form so that control of both attributes can be built into the process itself. In addition, SQC can instantly spot malfunctions and show where they occur—a worn tool, a dirty spray gun, an overheating furnace. And because it can do this with a small sample, malfunctions are reported almost immediately, allowing machine operators to correct problems in real time. Further, SQC quickly identifies the impact of any change on the performance of the entire process. (Indeed, in some applications developed by Deming's Japanese disciples, computers can simulate the effects of a proposed change in advance.) Finally, SQC identifies where, and often how, the quality and productivity of the entire process can be continuously improved. This used to be called the "Shewhart Cycle" and then the "Deming Cycle"; now it is *kaizen*, the Japanese term for continuous improvement.

But these engineering characteristics explain only a fraction of SQC's results. Above all, they do not explain the productivity gap between Japanese and U.S. factories. Even after adjusting for their far greater reliance on outside suppliers, Toyota, Honda, and Nissan turn out two or three times more cars per worker than comparable U.S. or European plants do. Building quality into the process accounts for no more than one-third of this difference. Japan's major productivity gains are the result of social changes brought about by SQC.

The Japanese employ proportionately more machine operators in direct production work than Ford or GM. In fact, the introduction of SQC almost always increases the number of machine operators. But this increase is offset many times over by the sharp drop in the number of nonoperators: inspectors, above all, but also the people who do not *do* but *fix*, like repair crews and "fire fighters" of all kinds.

In U.S. factories, especially mass-production plants, such nonoperating, blue-collar employees substantially outnumber operators. In some plants, the ratio is two to one. Few of these workers are needed under SQC. Moreover, first-line supervisors also are gradually eliminated, with only a handful of trainers taking their place. In other words, not only does SQC make it possible for machine operators to be in control of their work, it makes such control almost mandatory. No one else has the hands-on knowledge needed to act effectively on the information that SQC constantly feeds back.

By aligning information with accountability, SQC resolves a heretofore irresolvable conflict. For more than a century, two basic approaches to manufacturing have prevailed, especially in the United States. One is the engineering approach pioneered by Frederick Winslow Taylor's "scientific management." The other is the "human relations" (or "human resources") approach developed before World War I by Andrew Carnegie, Julius Rosenwald of Sears Roebuck, and Hugo Münsterberg, a Harvard psychologist. The two approaches have always been considered antitheses, indeed, mutually exclusive. In SQC, they come together.

Taylor and his disciples were just as determined as Deming to build quality and productivity into the manufacturing process. Taylor asserted that his "one right way" guaranteed zero-defects quality; he was as vehemently opposed to inspectors as Deming is today. So was Henry Ford, who claimed that his assembly line built quality and productivity into the process (though he was otherwise untouched by Taylor's scientific management and probably did not even know about it). But without SQC's rigorous methodology, neither scientific management nor the assembly line could actually deliver built-in process control. With all their successes, both scientific management and the assembly line had to fall back on massive inspection, to fix problems rather than eliminate them.

The human-relations approach sees the knowledge and pride of line workers as the greatest resource for controlling and improving quality and productivity. It too has had important successes. But without the kind of information SQC provides, you cannot readily distinguish productive activity from busy-ness. It is also hard to tell whether a proposed modification will truly improve the process or simply make things look better in one corner, only to make them worse overall.

Quality circles, which were actually invented and widely used in U.S. industry during World War II, have been successful in Japan because they came in after SQC had been established. As a result, both the quality circle and management have objective information about the effects of workers' suggestions. In contrast, most U.S. quality circles of the last 20 years have failed despite great enthusiasm, especially on the part of workers. The reason? They were established without SQC, that is, without rigorous and reliable feedback.

A good many U.S. manufacturers have built quality and productivity into their manufacturing processes without SQC and yet with a minimum of inspection and fixing. Johnson & Johnson is one such example. Other companies have successfully put machine operators in control of the manufacturing process without instituting SQC. IBM long ago replaced all first-line supervisors with a handful of "managers" whose main task is to train, while Herman Miller achieves zero-defects quality and high productivity through continuous training and productivity-sharing incentives.

But these are exceptions. In the main, the United States has lacked the methodology to build quality and productivity into the manufacturing process. Similarly, we have lacked the methodology to move responsibility for the process and control of it to the machine operator, to put into practice what the mathematician Norbert Wiener called the "human use of human beings."

SQC makes it possible to attain both traditional aspirations: high quality and productivity on the one hand, work worthy of human beings on the other. By fulfilling the aims of the traditional factory, it provides the capstone for the edifice of twentieth century manufacturing that Frederick Taylor and Henry Ford designed.

Bean counters do not enjoy a good press these days. They are blamed for all the ills that afflict U.S. manufacturing. But the bean counters will have the last laugh. In the factory of 1999, manufacturing accounting will play as big a role as it ever did and probably even a bigger one. But the beans will be counted differently. The new manufacturing accounting, which might more accurately be called "manufacturing economics," differs radically from traditional cost accounting in its basic concepts. Its aim is to integrate manufacturing with business strategy.

Manufacturing cost accounting (cost accounting's rarely used full name) is the third leg of the stool—the other legs being scientific management and the assembly line—on which modern manufacturing industry rests. Without cost accounting, these two could never have become fully effective. It too is American in origin. Developed in the 1920s by General Motors, General Electric, and Western Electric (AT&T's manufacturing arm), the new cost accounting, not technology, gave GM and GE the competitive edge that made them worldwide industry leaders. Following World War II, cost accounting became a major U.S. export.

But by that time, cost accounting's limitations also were becoming apparent. Four are particularly important. First, cost accounting is based on the realities of the 1920s, when direct, blue-collar labor accounted for 80% of all manufacturing costs other than raw materials. Consequently, cost accounting equates "cost" with direct labor costs. Everything else is "miscellaneous," lumped together as overhead.

These days, however, a plant in which direct labor costs run as high as 25% is a rare exception. Even in automobiles, the most labor intensive of the major industries, direct labor costs in up-to-date plants (such as those the Japanese are building in the United States and some of the new Ford plants) are down to 18%. And 8% to 12% is fast becoming the industrial norm. One large manufacturing company with a labor-intensive process, Beckman Instruments, now considers labor costs "miscellaneous." But typically, cost accounting systems are still based on direct labor costs that are carefully, indeed minutely, accounted for. The remaining costs—and that can mean 80% to 90%—are allocated by ratios that everyone knows are purely arbitrary and totally misleading: in direct proportion to a product's labor costs, for example, or to its dollar volume.

Second, the benefits of a change in process or in method are primarily defined in terms of labor cost savings. If other savings are considered at all, it is usually on the basis of the same arbitrary allocation by which costs other than direct labor are accounted for.

Even more serious is the third limitation, one built into the traditional cost accounting system. Like a sundial, which shows the hours when the sun shines but gives no information on a cloudy day or at night, traditional cost accounting measures only the costs of producing. It ignores the costs of nonproducing, whether they result from machine downtime or from quality defects that require scrapping or reworking a product or part.

Standard cost accounting assumes that the manufacturing process turns out good products 80% of the time. But we now know that even with the best SQC, nonproducing time consumes far more than 20% of total production time. In some plants, it accounts for 50%. And nonproducing time costs as much as producing time does—in wages, heat, lighting, interest, salaries, even raw materials. Yet the traditional system measures none of this.

Finally, manufacturing cost accounting assumes the factory is an isolated entity. Cost savings in the factory are "real." The rest is "speculation"—for example, the impact of a manufacturing process change on a product's acceptance in the market or on service quality. GM's plight since the 1970s illustrates the problem with this assumption. Marketing people were unhappy with top management's decision to build all car models, from Chevrolet to Cadillac, from the same small number of bodies, frames, and engines. But the cost accounting model showed that such commonality would produce substantial labor cost savings. And so marketing's argument that GM cars would lose customer appeal as they looked more and more alike was brushed aside as speculation. In effect, traditional cost accounting can hardly justify a product *improvement*, let alone a product or process *innovation*. Automation, for instance, shows up as a cost but almost never as a benefit.

All this we have known for close to 40 years. And for 30 years, accounting scholars, government accountants, industry accountants, and accounting firms have worked hard to reform the system. They have made substantial improvements. But since the reform attempts tried to build on the traditional system, the original limitations remain.

What triggered the change to the new manufacturing accounting was the frustration of factory-automation equipment makers. The potential users, the people in the plants, badly wanted the new equipment. But top management could not be persuaded to spend the money on numerically controlled machine tools or robots that could rapidly change tools, fixtures, and molds. The benefits of automated equipment, we now know, lie primarily in the reduction of nonproducing time by improving quality (that is, getting it right the first time) and by sharply curtailing machine downtime in changing over from one model or product to another. But these gains cost accounting does not document.

Out of this frustration came Computer-Aided Manufacturing-International, or CAM-I, a cooperative effort by automation producers, multinational manufacturers, and accountants to develop a new cost accounting system. Started in 1986, CAM-I is just beginning to influence manufacturing practice. But already it has unleashed an intellectual revolution. The most exciting and innovative work in management today is found in accounting theory, with new concepts, new approaches, new methodology—even what might be called new economic philosophy—rapidly taking shape. And while there is enormous controversy over specifics, the lineaments of the new manufacturing accounting are becoming clearer every day.

As soon as CAM-I began its work, it became apparent that the traditional accounting system could not be reformed. It had to be replaced. Labor costs are clearly the wrong unit of measurement in manufacturing. But—and this is a new insight—so are all the other elements of production. The new measurement unit has to be time. The costs for a given period of time must be assumed to be fixed; there are no "variable" costs. Even material costs are more fixed than variable, since defective output uses as much material as good output does. The only thing that is both variable and controllable is how much time a given process takes. And "benefit" is whatever reduces that time. In one fell swoop, this insight eliminates the first three of cost accounting's four traditional limitations.

But the new cost concepts go even further by redefining what costs and benefits really are. For example, in the traditional cost accounting system, finished-goods inventory costs nothing because it does not absorb any direct labor. It is treated as an "asset." In the new manufacturing accounting, however, inventory of finished goods is a "sunk cost" (an economist's, not an accountant's, term). Stuff that sits in inventory does not earn anything. In fact, it ties down expensive money and absorbs time. As a result, its time costs are high. The new accounting measures these time costs against the benefits of finished-goods inventory (quicker customer service, for instance).

Yet manufacturing accounting still faces the challenge of eliminating the fourth limitation of traditional cost accounting: its inability to bring into the measurement of factory performance the impact of manufacturing changes on the total business—the return in the marketplace of an investment in automation, for instance, or the risk in not making an investment that would speed up production change-overs. The in-plant costs and benefits of such decisions can now be worked out with considerable accuracy. But the business consequences are indeed speculative. One can only say, "Surely, this should help us get more sales," or "If we don't do this, we risk falling behind in customer service." But how do you quantify such opinions?

Cost accounting's strength has always been that it confines itself to the measurable and thus gives objective answers. But if intangibles are brought into its equations, cost accounting will only raise more questions. How to proceed is thus hotly debated, and with good reason. Still, everyone agrees that these business impacts have to be integrated into the measurement of factory performance, that is, into manufacturing accounting. One way or another, the new accounting will force managers, both inside and outside the plant, to make manufacturing decisions as *business* decisions.

Henry Ford's epigram, "The customer can have any color as long as it's black," has entered American folklore. But few people realize what Ford meant: flexibility costs time and money, and the customer won't pay for it. Even fewer people realize that in the mid-1920s, the "new" cost accounting made it possible for GM to beat Ford by giving customers both colors and annual model changes at no additional cost.

By now, most manufacturers can do what GM learned to do roughly 70 years ago. Indeed, many go quite a bit further in combining standardization with flexibility. They can, for example, build a variety of end products from a fairly small number of standardized parts. Still, manufacturing people tend to think like Henry Ford: you can have either standardization at low cost or flexibility at high cost, but not both.

The factory of 1999, however, will be based on the premise that you not only *can* have both but also *must* have both—and at low cost. But to achieve this, the factory will have to be structured quite differently.

Today's factory is a battleship. The plant of 1999 will be a "flotilla," consisting of modules centered either around a stage in the production process or around a number of closely related operations. Though overall command and control will still exist, each module will have its own command and control. And each, like the ships in a flotilla, will be maneuverable, both in terms of its position in the entire process and its relationship to other modules. This organization will give each module the benefits of standardization and, at the same time, give the whole process greater flexibility. Thus it will allow rapid changes in design and product, rapid response to market demands, and low-cost production of "options" or "specials" in fairly small batches.

No such plant exists today. No one can yet build it. But many manufacturers, large and small, are moving toward the flotilla structure: among them are some of Westinghouse's U.S. plants, Asea Brown Boveri's robotics plant in Sweden, and several large printing plants, especially in Japan.

The biggest impetus for this development probably came from GM's failure to get a return on its massive (at least $30 billion and perhaps $40 billion) investment in automation. GM, it seems, used the new machines to improve its existing process, that is, to make the assembly line more efficient. But the process instead became less flexible and less able to accomplish rapid change.

Meanwhile, Japanese automakers and Ford were spending less and attaining more flexibility. In these plants, the line still exists, but it is discontinuous rather than tightly tied together. The new equipment is being used to speed changes, for example, automating changeovers of jigs, tools, and fixtures. So the line has acquired a good bit of the flexibility of traditional batch production without losing its standardization. Standardization and flexibility are thus no longer an either-or proposition. They are—as indeed they must be—melded together.

This means a different balance between standardization and flexibility, however, for different parts of the manufacturing process. An "average" balance across the plant will do nothing very well. If imposed throughout the line, it will simply result in high rigidity and big costs for the entire process, which is apparently what happened at GM. What is required is a reorganization of the process into modules, each with its own optimal balance.

Moreover, the relationships between these modules may have to change whenever the product, process, or distribution changes. Switching from selling heavy equipment to leasing it, for instance, may drastically change the ratio between finished-product output and spare-parts output. Or a fairly minor model change may alter the sequence in which major parts are assembled into the finished product. There is nothing very new in this, of course. But under the traditional line structure, such changes are ignored, or they take forever to accomplish. With competition intensifying and product life cycles shortening all the time, such changes cannot be ignored, and they have to be done fast. Hence the flotilla's modular organization.

But this organization requires more than a fairly drastic change in the factory's physical structure. It requires, above all, different communication and information. In the traditional plant, each sector and department reports separately upstairs. And it reports what upstairs has asked for. In the factory of 1999, sectors and departments will have to think through what information they owe to whom and what information they need from whom. A good deal of this information will flow sideways and across department lines, not upstairs. The factory of 1999 will be an information network.

Consequently, all the managers in a plant will have to know and understand the entire process, just as the destroyer commander has to know and understand the tactical plan of the entire flotilla. In the factory of 1999, managers will have to think and act as team members, mindful of the performance of the whole. Above all, they will have to ask: What do the people running the other modules need to know about the characteristics, the capacity, the plans, and the performance of my unit? And what, in turn, do we in my module need to know about theirs?

The last of the new concepts transforming manufacturing is systems design, in which the whole of manufacturing is seen as an integrated process that converts materials into goods, that is, into economic satisfactions.

Marks & Spencer, the British retail chain, designed the first such system in the 1930s. Marks & Spencer designs and tests the goods (whether textiles or foods) it has decided to sell. It designates one manufacturer to make each product under contract. It works with the manufacturer to produce the right merchandise with the right quality at the right price. Finally, it organizes just-in-time delivery of the finished products to its stores. The entire process is governed by a meticulous forecast as to when the goods will move off store shelves and into customers' shopping bags. In the last ten years or so, such systems management has become common in retailing.

Though systems organization is still rare in manufacturing, it was actually first attempted there. In the early 1920s, when the Model T was in its full glory, Henry Ford decided to control the entire process of making and moving all the supplies and parts needed by his new plant, the gigantic River Rouge. He built his own steel mill and glass plant. He founded plantations in Brazil to grow rubber for tires. He bought the railroad that brought supplies to River Rouge and carried away the finished cars. He even toyed with the idea of building his own service centers nationwide and staffing them with mechanics trained in Ford-owned schools. But Ford conceived of all this as a financial edifice held together by ownership. Instead of building a system, he built a conglomerate, an unwieldy monster that was expensive, unmanageable, and horrendously unprofitable.

In contrast, the new manufacturing system is not "controlled" at all. Most of its parts are independent—independent suppliers at one end, customers at the other. Nor is it plant centered, as Ford's organization was. The new system sees the plant as little more than a wide place in the manufacturing stream. Planning and scheduling start with shipment to the final customer, just as they do at Marks & Spencer. Delays, halts, and redundancies have to be designed into the system—a warehouse here, an extra supply of parts and tools there, a stock of old products that are no longer being made but are still occasionally demanded by the market. These are necessary imperfections in a continuous flow that is governed and directed by information.

What has pushed American manufacturers into such systems design is the trouble they encountered when they copied Japan's just-in-time methods for supplying plants with materials and parts. The trouble could have been predicted, for the Japanese scheme is founded in social and logistic conditions unique to that country and unknown in the United States. Yet the shift seemed to American manufacturers a matter of procedure, indeed, almost trivial. Company after company found, however, that just-in-time delivery of supplies and parts created turbulence throughout their plants. And while no one could figure out what the problem was, the one thing that became clear was that with just-in-time deliveries, the plant no longer functions as a step-by-step process that begins at the receiving dock and ends when finished goods move into the shipping room. Instead, the plant must be redesigned from the end backwards and managed as an integrated flow.

Manufacturing experts, executives, and professors have urged such an approach for two or three decades now. And some industries, such as petroleum refining and large-scale construction, do practice it. But by and large, American and European manufacturing plants are neither systems designed nor systems managed. In fact, few companies have enough knowledge about what goes on in their plants to run them as systems. Just-in-time delivery, however, forces managers to ask systems questions: Where in the plant do we need redundancy? Where should we place the burden of adjustments? What costs should we incur in one place to minimize delay, risk, and vulnerability in another?

A few companies are even beginning to extend the systems concept of manufacturing beyond the plant and into the marketplace. Caterpillar, for instance, organizes its manufacturing to supply any replacement part anywhere in the world within 48 hours. But companies like this are still exceptions; they must become the rule. As soon as we define manufacturing as the process that converts things into economic satisfactions, it becomes clear that producing does not stop when the product leaves the factory. Physical distribution and product service are still part of the production process and should be integrated with it, coordinated with it, managed together with it. It is already widely recognized that servicing the product must be a major consideration during its design and production. By 1999, systems manufacturing will have an increasing influence on how we design and remodel plants and on how we manage manufacturing businesses.

Traditionally, manufacturing businesses have been organized "in series," with functions such as engineering, manufacturing, and marketing as successive steps. These days, that system is often complemented by a parallel team organization (Procter & Gamble's product management teams are a well-known example), which brings various functions together from the inception of a new product or process project. If manufacturing is a system, however, every decision in a manufacturing business becomes a manufacturing decision. Every decision should meet manufacturing's requirements and needs and in turn should exploit the strengths and capabilities of a company's particular manufacturing system.

When Honda decided six or seven years ago to make a new, upscale car for the U.S. market, the most heated strategic debate was not about design, performance, or price. It was about whether to distribute the Acura through Honda's well-established dealer network or to create a new market segment by building separate Acura dealerships at high cost and risk. This was a marketing issue, of course. But the decision was made by a team of design, engineering, manufacturing, and marketing people. And what tilted the balance toward the separate dealer network was a manufacturing consideration: the design for which independent distribution and service made most sense was the design that best utilized Honda's manufacturing capabilities.

Full realization of the systems concept in manufacturing is years away. It may not require a new Henry Ford. But it will certainly require very different management and very different managers. Every manager in tomorrow's manufacturing business will have to know and understand the manufacturing system. We might well adopt the Japanese custom of starting all new management people in the plant and in manufacturing jobs for the first few years of their careers. Indeed, we might go even further and require managers throughout the company to rotate into factory assignments throughout their careers—just as army officers return regularly to troop duty.

In the new manufacturing business, manufacturing is the integrator that ties everything together. It creates the economic value that pays for everything and everybody. Thus the greatest impact of the manufacturing systems concept will not be on the production process. As with SQC, its greatest impact will be on social and human concerns—on career ladders, for instance, or more important, on the transformation of *functional* managers into *business* managers, each with a specific role, but all members of the same production and the same cast. And surely, the manufacturing businesses of tomorrow will not be run by financial executives, marketers, or lawyers inexperienced in manufacturing, as so many U.S. companies are today.

There are important differences among these four concepts. Consider, for instance, what each means by "the factory." In SQC, the factory is a place where people work. In management accounting and the flotilla concept of flexible manufacturing, it is a place where work is being done—it makes no difference whether by people, by white mice, or by robots. In the systems concept, the factory is not a place at all; it is a stage in a process that adds economic value to materials. In theory, at least, the factory cannot and certainly should not be designed, let alone built, until the entire process of "making"—all the way to the final customer—is understood. Thus defining the factory is much more than a theoretical or semantic exercise. It has immediate practical consequences on plant design, location, and size; on what activities are to be brought together in one manufacturing complex; even on how much and in what to invest.

Similarly, each of these concepts reflects a particular mind-set. To apply SQC, you don't have to think, you have to do. Management accounting concentrates on technical analysis, while the flotilla concept focuses on organization design and work flow. In the systems concept, there is great temptation to keep on thinking and never get to the doing. Each concept has its own tools, its own language, and addresses different people.

Nevertheless, what these four concepts have in common is far more important than their differences. Nowhere is this more apparent than in their assumption that the manufacturing process is a configuration, a whole that is greater than the sum of its parts. Traditional approaches all see the factory as a collection of individual machines and individual operations. The nineteenth century factory was an assemblage of machines. Taylor's scientific management broke up each job into individual operations and then put those operations together into new and different jobs. "Modern" twentieth century concepts—the assembly line and cost accounting—define performance as the sum of lowest cost operations. But none of the new concepts is much concerned with performance of the parts. Indeed, the parts as such can only underperform. The process produces results.

Management also will reflect this new perspective. SQC is the most nearly conventional in its implications for managers, since it does not so much change their job as shift much of it to the work force. But even managers with no business responsibility (and under SQC, plant people have none) will have to manage with an awareness of business considerations well beyond the plant. And every manufacturing manager will be responsible for integrating people, materials, machines, and time. Thus every manufacturing manager ten years hence will have to learn and practice a discipline that integrates engineering, management of people, and business economics into the manufacturing process. Quite a few manufacturing people are doing this already, of course—though usually unaware that they are doing something new and different. Yet such a discipline has not been systematized and is still not taught in engineering schools or business schools.

These four concepts are synergistic in the best sense of this much-abused term. Together—but only together—they tackle the conflicts that have most troubled traditional, twentieth century mass-production plants: the conflicts between people and machines, time and money, standardization and flexibility, and functions and systems. The key is that every one of these concepts defines performance as productivity and conceives of manufacturing as the physical process that adds economic value to materials. Each tries to provide economic value in a different way. But they share the same theory of manufacturing.

2
Marketing in an Age of Diversity

Regis McKenna

Spreading east from California, a new individualism has taken root across the United States. Gone is the convenient fiction of a single, homogeneous market. The days of a uniformly accepted view of the world are over. Today diversity exerts tremendous influence, both economically and politically.

Technology and social change are interdependent. Companies are using new flexible technology, like computer-aided design and manufacturing and software customization, to create astonishing diversity in the marketplace and society. And individuals temporarily coalescing into "micromajorities" are making use of platforms—media, education, and the law—to express their desires.

In the marketing world, for example, the protests of thousands of consumers, broadcast by the media as an event of cultural significance, were enough to force Coca-Cola to reverse its decision to do away with "classic" Coke. On the political scene, vociferous minorities, sophisticated in using communication technology, exert influence greatly disproportionate to their numbers: the Moral Majority is really just another minority—but focused and amplified. When we see wealthy people driving Volkswagens and pickup trucks, it is clear that this is a society where individual tastes are no longer predictable; marketers cannot easily and neatly categorize their customer base.

Since the early 1970s, new technology has spawned products aimed at diverse, new sectors and market niches. Computer-aided technologies now allow companies to customize virtually any product, from designer jeans to designer genes, serving ever narrower customer

Exhibit 2-1   Supermarket Items

Today's consumers have 60% more product variety in the supermarket than they did in 1981.

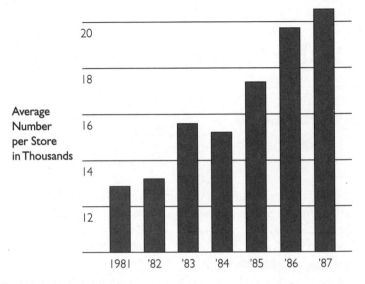

Source: Food Marketing Institute

needs. With this newfound technology, manufacturers are making more and more high-quality products in smaller and smaller batches; today 75% of all machined parts are produced in batches of 50 or fewer.

Consumers demand—and get—more variety and options in all kinds of products, from cars to clothes. Auto buyers, for example, can choose from 300 different types of cars and light trucks, domestic and imported, and get variations within each of those lines. Beer drinkers now have 400 brands to sample. The number of products in super-markets soared from 13,000 in 1981 to 21,000 in 1987. (See Exhibit 2-1 "Supermarket Items.") There are so many new items that stores can demand hefty fees from packaged-foods manufacturers just for dis-playing new items on grocery shelves.

Deregulation has also increased the number of choices—from a flurry of competing airfares to automated banking to single-premium life insurance you can buy at Sears. The government has even adapted antitrust laws to permit companies to serve emerging micromarkets: the Orphan Drug Act of 1983, for instance, gave pharmaceutical companies tax breaks and a seven-year monopoly on any drugs that serve fewer than 200,000 people.

Diversity and niches create tough problems for old-line companies more accustomed to mass markets. Sears, the country's largest retailer, is trying to reposition its products, which traditionally have appealed to older middle-class and blue-collar customers. To lure younger, style-minded buyers, Sears has come up with celebrity-signature lines, fashion boutiques, and a new line of children's clothing, McKids, playing off the McDonald's draw. New, smaller stores, specialty catalogs, and merchandise tailored to regional tastes are all part of Sears's effort to reach a new clientele—without alienating its old one.

Faced with slimmer profits from staples like detergents, diapers, and toothpaste, and lackluster results from new food and beverage products, Procter & Gamble, the world's largest marketer, is rethinking what it should sell and how to sell it. The company is now concentrating on health products; it has high hopes for a fat substitute called "olestra," which may take some of the junk out of junk food. At the same time that P&G is shifting its product thinking, it also is changing its organization, opening up and streamlining its highly insular pyramidal management structure as part of a larger effort to listen and respond to customers. Small groups that include both factory workers and executives work on cutting costs, while other teams look for new ways to speed products to market.

In trying to respond to the new demands of a diverse market, the problem that giants like Sears and P&G face is not fundamental change, not a total turnabout in what an entire nation of consumers wants. Rather, it is the fracturing of mass markets. To contend with diversity, managers must drastically alter how they design, manufacture, market, and sell their products. Marketing in the age of diversity means:

- More options for goods producers and more choices for consumers.
- Less perceived differentiation among similar products.
- Intensified competition, with promotional efforts sounding more and more alike, approaching "white noise" in the marketplace.
- Newly minted meanings for words and phrases as marketers try to "invent" differentiation.

- Disposable information as consumers try to cope with information deluge from print, television, computer terminal, telephone, fax, satellite dish.
- Customization by users as flexible manufacturing makes niche production every bit as economic as mass production.
- Changing leverage criteria as economies of scale give way to economies of knowledge—knowledge of the customer's business, of current and likely future technology trends, and of the competitive environment that allows the rapid development of new products and services.
- Changing company structure as large corporations continue to downsize to compete with smaller niche players that nibble at their markets.
- Smaller wins—fewer chances for gigantic wins in mass markets, but more opportunities for healthy profits in smaller markets.

The Decline of Branding, the Rise of "Other"

In today's fractured marketplace, tried-and-true marketing techniques from the past no longer work for most products—particularly for complex ones based on new technology. Branding products and seizing market share, for instance, no longer guarantee loyal customers. In one case after another, the old, established brands have been supplanted by the rise of "other."

Television viewers in 1983 and 1984, for instance, tuned out the big three broadcasters to watch cable and independent "narrow cast" stations. The trend continued in 1987 as the big three networks lost 9% of their viewers—more than six million people. Small companies appealing to niche-oriented viewers attacked the majority market share. NBC responded by buying a cable television company for $20 million.

No single brand can claim the largest share of the gate array, integrated circuit, or computer market. Even IBM has lost its reign over the personal computer field—not to one fast-charging competitor but to an assortment of smaller producers. Tropicana, Minute Maid, and Citrus Hill actually account for less than half the frozen orange juice market. A full 56% belongs to hundreds of mostly small private labels. (See Exhibit 2-2 "Frozen Concentrate.") In one area after another, "other" has become the major market holder.

IBM's story of lost market share bears elaboration, in large part because of the company's almost legendary position in the U.S. busi-

Exhibit 2-2   Frozen Concentrate

The "big" brands aren't so big after all, "Other" is the market leader.

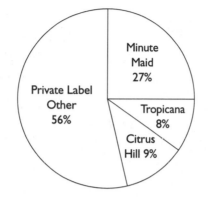

Data: Coca-Cola Foods BCI Holdings, A.C. Nielsen Co.

ness pantheon. After its rise in the personal computer market through 1984 (see Exhibit 2-3 "PC Market Share, 1984."), IBM found its stronghold eroding—but not to just one huge competitor that could be identified and stalked methodically. IBM could no longer rely on tracking the dozen or so companies that had been its steady competition for almost two decades. Instead, more than 300 clone producers worldwide intruded on Big Blue's territory. (See Exhibit 2-4 "PC Market Share, 1986.") Moreover, IBM has faced the same competitive challenge in one product area after another, from supercomputers to networks. In response, IBM has changed how it does business. In the past, IBM wouldn't even bother to enter a market lacking a value of at least $100 million. But today, as customer groups diversify and markets splinter, that criterion is obsolete. The shift in competition has also prompted IBM to reorganize, decentralizing the company into five autonomous groups so decisions can be made closer to customers.

Similar stories abound in other industries. Kodak dominated film processing in the United States until little kiosks sprang up in shopping centers and ate up that market. In the late 1960s, the U.S. semiconductor industry consisted of 100 companies; today there are more than 300. In fact, practically every industry has more of every kind of company catering to the consumer's love of diversity—more ice cream companies, more cookie companies, more weight loss and exercise

Exhibit 2-3   PC Market Share, 1984 (By Factory Revenue, U.S. only)

IBM led the market then . . .

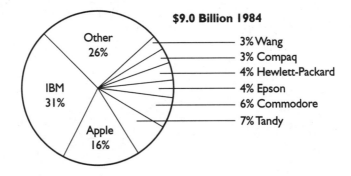

$9.0 Billion 1984

- Other 26%
- IBM 31%
- Apple 16%
- 3% Wang
- 3% Compaq
- 4% Hewlett-Packard
- 4% Epson
- 6% Commodore
- 7% Tandy

Source: Future Computing, Inc. 1986

Exhibit 2-4   PC Market Share, 1986 (By Factory Revenue, U.S. only)

. . . but now it's been eclipsed by clone companies it can't identify or stalk.

$14.5 Billion 1986

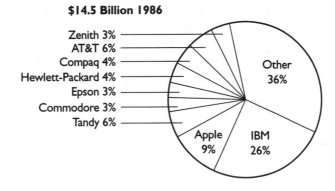

- Zenith 3%
- AT&T 6%
- Compaq 4%
- Hewlett-Packard 4%
- Epson 3%
- Commodore 3%
- Tandy 6%
- Other 36%
- Apple 9%
- IBM 26%

Source: Future Computing, Inc. 1986

companies. In 1987, enterprising managers started 233,000 new businesses of all types to offer customers their choice of "other."

The False Security of Market Share

The proliferation of successful small companies dramatizes how the security of majority market share—seized by a large corporation and held unchallenged for decades—is now a dangerous anachronism. In the past, the dominant marketing models drew on the measurement and control notions embedded in engineering and manufacturing. The underlying mechanistic logic was that companies could measure everything, and anything they could measure, they could control—including customers. Market-share measurements became a way to understand the marketplace and thus to control it. For example, marketers used to be able to pin down a target customer with relative ease: if it were a man, he was between 25 and 35 years old, married, with two-and-a-half children, and half a dog. Since he was one of so many measurable men in a mass society, marketers assumed that they could manipulate the market just by knowing the demographic characteristics.

But we don't live in that world anymore, and those kinds of measurements are meaningless. Marketers trying to measure that same "ideal" customer today would discover that the pattern no longer holds; that married fellow with two-and-a-half kids could now be divorced, situated in New York instead of Minnesota, and living in a condo instead of a brick colonial. These days, the idea of market share is a trap that can lull businesspeople into a false sense of security.

Managers should wake up every morning uncertain about the marketplace, because it is invariably changing. That's why five-year plans are dangerous: Who can pinpoint what the market will be five years from now? The president of one large industrial corporation recently told me, "The only thing we know about our business plan is that it's wrong. It's either too high or too low—but we never know which."

In the old days, mass marketing offered an easy solution: "just run some ads." Not today. IBM tried that approach with the PC Jr., laying out an estimated $100 million on advertising—before the product failed. AT&T spent tens of millions of dollars running ads for its computer products.

In sharp contrast, Digital Equipment Corporation spent very little on expensive national television advertising and managed to wrest a healthy market position. Skipping the expensive mass-advertising campaigns, DEC concentrated on developing its reputation in the computer business by solving problems for niche markets. Word of mouth sold DEC products. The company focused its marketing and sales staffs where they already had business and aimed its message at people who actually make the decision on what machines to buy. DEC clearly understood that no one buys a complex product like a computer without a reliable outside reference—however elaborate the company's promotion.

Niche Marketing: Selling Big by Selling Small

Intel was in the personal computer business two years before Apple started in Steve Jobs's garage. The company produced the first microprocessor chip and subsequently developed an early version of what became known as the hobby computer, sold in electronics hobby stores. An early Intel advertisement in *Scientific American* showed a junior high school student using the product. Intel's market research, however, revealed that the market for hobbyists was quite small and it abandoned the project. Two years later, Apple built itself on the hobbyist market. As it turned out, many of the early users of personal computers in education, small business, and the professional markets were hobbyists or enthusiasts.

I recently looked at several market forecasts made by research organizations in 1978 projecting the size of the personal computer market in 1985. The most optimistic forecast looked for a $2 billion market. It exceeded $25 billion.

Most large markets evolve from niche markets. That's because niche marketing teaches many important lessons about customers—in particular, to think of customers as individuals and to respond to their special needs. Niche marketing depends on word-of-mouth references and infrastructure development, a broadening of people in related industries whose opinions are crucial to the product's success.

Infrastructure marketing can be applied to almost all markets. In the medical area, for example, recognized research gurus in a given field—diabetes, cancer, heart disease—will first experiment with new devices or drugs at research institutions. Universities and research institutions become identified by their specialties. Experts in a particular area talk to each other, read the same journals, and attend the same conferences. Many companies form their own scientific advisory boards designed to tap into the members' expertise and to build credibility for new technology and products. The word of mouth created by infrastructure marketing can make or break a new drug or a new supplier. Conductus, a superconductor company in Palo Alto, built its business around an advisory board of seven top scientists from Stanford University and Berkeley.

Represented graphically, infrastructure development would look like an inverted pyramid. So Apple's pyramid, for instance, would include the references of influential users, software designers who create programs, dealers, industry consultants, analysts, the press, and, most important, customers.

Customer focus derived from niche marketing helps companies respond faster to demand changes. That is the meaning of today's most critical requirement—that companies become market driven. From the board of directors down through the ranks, company leaders must educate everyone to the singular importance of the customer, who is no longer a faceless, abstract entity or a mass statistic.

Because niche markets are not easily identified in their infancy, managers must keep one foot in the technology to know its potential and one foot in the market to see opportunity. Tandem Computers built its solid customer base by adapting its products to the emerging on-line transaction market. Jimmy Treybig, president and CEO, told me that the company had to learn the market's language. Bankers don't talk about MIPS (millions of instructions per second) the way computer people do, he said; they talk about transactions. So Tandem built its products and marketing position to become the leading computer in the transaction market. Not long before, Treybig said, he had been on a nation-wide tour visiting key customers. "Guess who was calling on my customers just a few days ahead of me? John Akers"—chairman of IBM.

Many electronics companies have developed teams consisting of software and hardware development engineers, quality control and manufacturing people, as well as marketing and salespeople—who all visit customers or play key roles in dealing with customers. Convex Computer and Tandem use this approach. Whatever method a company may use, the purpose is the same: to get the entire company to focus on the fragmented, ever-evolving customer base as if it were an integral part of the organization.

The Integrated Product

Competition from small companies in fractured markets has even produced dramatic changes in how companies define their products. The product is no longer just the thing itself; it includes service, word-of-mouth references, company financial reports, the technology, and even the personal image of the CEO.

As a result, product marketing and service marketing, formerly two distinct fields, have become a single hybrid. For example, Genentech, which manufactures a growth hormone, arms its sales force with laptop computers. When a Genentech salesperson visits an endocrinologist, the physician can tie into a data base of all the tests run on people with characteristics similar to his or her patients. The computer represents an extended set of services married to the original product.

Or take the example of Apple Computer and Quantum Corporation. These firms entered a joint venture offering on-line interactive computer services for Apple computer users. In addition to a long list of transaction services that reads like a television programming guide, Apple product service, support, and even simple maintenance have been integrated into the product itself. Prodigy, a joint venture between IBM and Sears, offers IBM and Apple users access to banking, shopping, the stock market, regional weather forecasts, sports statistics, and encyclopedias of all kinds—and even direct advice from Sylvia Porter, Howard Cosell, and Ask Beth.

In consumer products, service has become the predominant distinguishing feature. Lands' End promotes its catalog-marketed outdoorsy clothes by guaranteeing products unconditionally and promising to ship orders within 24 to 48 hours. Carport, near Atlanta, offers air travelers an ultradeluxe parking service: it drives customers to their gates, checks their bags, and, while they are airborne, services, washes, and waxes their cars. "Macy's by Appointment" is a free shopping service for customers who are too busy or too baffled to make their own selections.

With so much choice backed by service, customers can afford to be fickle. As a result, references have become vital to product marketing.

And the more complex the product, the more complex the supporting references. After all, customers who switch toothpaste risk losing only a dollar or so if the new choice is a dud. But consumers buying a complete phone system or a computer system—or any other costly, long-term, and pervasive product—cannot afford to take their investments lightly. References become a part of the product, and they come in all kinds of forms. Company financial reports are a kind of reference. A person shopping for an expensive computer wants to see how profitable the company is; how can the company promise maintenance service if it's about to fold? Even the CEO's personality can make a sale. Customers who see Don Peterson of Ford splashed across a magazine cover—or Apple's John Sculley or Hewlett Packard's John Young—feel assured that a real person stands behind the complex and expensive product.

In this complicated world, customers weigh all these factors to winnow out the products they want from those they don't. Now more than ever, marketers must sell every aspect of their businesses as important elements of the products themselves.

The Customer As Customizer

Customer involvement in product design has become an accepted part of the development and marketing processes in many industries. In technologically driven products, which often evolve slowly as discoveries percolate to the surface, the customer can practically invent the market for a company.

Apple's experience with desktop publishing shows how companies and customers work together to create new applications—and new markets. Apple entered the field with the Macintosh personal computer, which offered good graphics and easy-to-use features. But desktop publishing didn't even exist then; it wasn't on anyone's pie chart as a defined market niche, and no one had predicted its emergence.

Apple's customers made it happen; newspapers and research organizations simply started using Macintosh's unique graphics capability to create charts and graphs. Early users made do with primitive software and printers, but that was enough to spark the imagination of other developers. Other hardware and software companies began developing products that could be combined with the Macintosh to enhance the user's publishing power. By visiting and talking to customers and other players in the marketplace, Apple began to realize desktop publishing's potential.

As customers explored the possibilities presented by the technology, the technology, in turn, developed to fit the customers' needs. The improved software evolved from a dynamic working relationship between company and customers, not from a rigid, bureaucratic headquarters determination of where Apple could find an extra slice of the marketing pie.

Technological innovation makes it easier to involve customers in design. For example, Milliken, the textile manufacturer, provides customers with computer terminals where they can select their own carpet designs from thousands of colors and patterns. Electronics customers, too, have assumed the role of product designer. New design tools allow companies like Tandem and Convex to design their own specialty chips, which the integrated-circuit suppliers then manufacture according to their specifications. Similarly, American Airlines designs its own computer systems. In cases like these, the design and manufacturing processes have been completely separated. So semiconductor companies—and many computer companies—have become raw-materials producers, with integration occurring all the way up the supply line.

The fact that customers have taken charge of design opens the door for value-added resellers, who integrate different materials and processes. These people are the essence of new-age marketers: they add value by understanding what happens in a doctor's office or a travel agency or a machine-tool plant and customize that service or product to the customer's needs. To capitalize on market changes, companies should follow these examples and work directly with customers—even before products hit the drawing boards.

The Evolution of Distribution

It's nearly impossible to make a prediction on the basis of past patterns. Perhaps many big institutions founded on assumptions of mass marketing and market share will disappear like dinosaurs. Or they'll evolve into closely integrated service and distribution organizations.

In fact, tremendous innovation in distribution channels has already begun in nearly every industry. Distribution channels have to be flexible to survive. As more flows into them, they have to change.

Grocery stores sell flowers and cameras. Convenience stores rent out videos. And television offers viewers direct purchasing access to everything from diamonds to snow blowers to a decent funeral.

To get products closer to customers, marketers are distributing more and more samples in more ways. Today laundry detergent arrives in the mail, magazines enfold perfume-doused tear-outs, and department stores offer chocolate samples. Software companies bind floppy disk samples into magazines or mail out diskettes that work only until a certain date, giving customers the chance to test a product before buying.

Every successful computer retailer has not only a showroom but also a classroom. The large computer retailers are not selling just to off-the-street traffic. Most of their volume now comes from a direct sales force calling on corporate America. In addition, all have application-development labs, extensive user-training programs and service centers—and some have recently experimented with private labeling their own computer product brands. The electronics community talks more and more about design centers—places where customers can get help customizing products and applications.

Today the product is an experience. As customers use it, they grow to trust it—and distribution represents the beginning of that evolving relationship. That's why computer companies donate their systems to elementary schools: schools are now a distribution channel for product experience.

Goliath plus David

Besides making changes in distribution channels, big corporations will also have to forge new partnerships with smaller companies. IBM, for example, already has ties to 1,500 small computer-service companies nationwide, offering help for IBM midsized machine owners. Olivetti makes personal computers for AT&T. All over the world, manufacturers are producing generic computer platforms; larger companies buy these, then add their own service-oriented, value-adding applications.

This approach seems almost inevitable considering what we know about patterns of research and development. Technological developments typically originate with basic research, move to applied research, to development, then to manufacturing and marketing. Very few U.S. companies do basic research; universities and various public and private labs generally shoulder that burden. Many big companies do applied R&D, while small companies concentrate on development. Basic and applied research means time and money. Consider the cases of two seminal inventions—antibiotics and television—the first of which took 30 years and the second 63 years from idea to the market.

Perhaps because of their narrow focus, small companies realize more development breakthroughs than larger ones. For example, the origins of recombinant DNA technology go back to the mid-1950s; it took Genentech only about six years to bring the world's first recombinant DNA commercial product to market.

A 1986 study by the Small Business Administration showed that 55% of innovations have come from companies with fewer than 500 employees, and twice as many innovations per employee come from small companies than from large ones. This finding, however, does not indicate that large companies are completely ineffective developers. Rather, the data suggest that small, venture-capitalized companies will scramble to invent a product that the market does not yet want, need, or perhaps even recognize; big companies will wait patiently for the market to develop so they can enter later with their strong manufacturing and marketing organizations.

The Japanese have shown us that it's wise to let small companies handle development—but only if large companies can somehow share that wisdom before the product reaches the market. From 1950 to 1978, Japanese companies held 32,000 licensing agreements to acquire foreign technology—mostly from the United States—for about $9 billion. In essence, the Japanese simply subcontracted out for R&D—and then used that investment in U.S. knowledge to dominate one market after another.

If orchestrated properly, agreements between large and small companies can prove mutually beneficial. When Genentech developed its first product, recombinant DNA insulin, the company chose not to compete against Eli Lilly, which held over 70% of the insulin market. Instead, Genentech entered into a licensing agreement with Lilly that put the larger company in charge of manufacturing and marketing the products developed by the smaller company. Over time, Genentech built its own manufacturing company while maintaining its proprietary product.

This model worked so well that it has shaped the fortunes of Silicon Valley. Of the 3,000 companies there, only a dozen hold places on the lists of America's largest corporations. Most of the companies are small developers of new products. Like the Japanese, large U.S. companies are now subcontracting development to these mostly high-tech startups. In the process, they are securing a critical resource—an ongoing relationship with a small, innovative enterprise.

Giant companies can compete in the newly diversifying markets if they recognize the importance of relationships—with small companies, within their own organizations, with their customers. Becoming market driven means abandoning old-style market-share thinking and instead tying the uniqueness of any product to the unique needs of the customer. This approach to marketing demands a revolution in how businesspeople act—and, even more important, in how they think. These changes are critical to success, but they can come only gradually, as managers and organizations adapt to the new rules of marketing in the age of diversity. As any good marketer knows, even instant success takes time.

PART

II

Efficiently Serving Customers Uniquely

3
Managing in an Age of Modularity

Carliss Y. Baldwin and Kim B. Clark

In the nineteenth century, railroads fundamentally altered the competitive landscape of business. By providing fast and cheap transportation, they forced previously protected regional companies into battles with distant rivals. The railroad companies also devised management practices to deal with their own complexity and high fixed costs that deeply influenced the second wave of industrialization at the turn of the century.

Today the computer industry is in a similar leading position. Not only have computer companies transformed a wide range of markets by introducing cheap and fast information processing, but they have also led the way toward a new industry structure that makes the best use of these processing abilities. At the heart of their remarkable advance is modularity—building a complex product or process from smaller subsystems that can be designed independently yet function together as a whole. Through the widespread adoption of modular designs, the computer industry has dramatically increased its rate of innovation. Indeed, it is modularity, more than speedy processing and communication or any other technology, that is responsible for the heightened pace of change that managers in the computer industry now face. And strategies based on modularity are the best way to deal with that change.

Many industries have long had a degree of modularity in their production processes. But a growing number of them are now poised to extend modularity to the design stage. Although they may have difficulty taking modularity as far as the computer industry has, managers in many industries stand to learn much about ways to employ this new approach from the experiences of their counterparts in computers.

A Solution to Growing Complexity

The popular and business presses have made much of the awesome power of computer technology. Storage capacities and processing speeds have skyrocketed while costs have remained the same or have fallen. These improvements have depended on enormous growth in the complexity of the product. The modern computer is a bewildering array of elements working in concert, evolving rapidly in precise and elaborate ways.

Modularity has enabled companies to handle this increasingly complex technology. By breaking up a product into subsystems, or *modules,* designers, producers, and users have gained enormous flexibility. Different companies can take responsibility for separate modules and be confident that a reliable product will arise from their collective efforts.

The first modular computer, the System/360, which IBM announced in 1964, effectively illustrates this approach. The designs of previous models from IBM and other mainframe manufacturers were unique; each had its own operating system, processor, peripherals, and application software. Every time a manufacturer introduced a new computer system to take advantage of improved technology, it had to develop software and components specifically for that system while continuing to maintain those for the previous systems. When end users switched to new machines, they had to rewrite all their existing programs, and they ran the risk of losing critical data if software conversions were botched. As a result, many customers were reluctant to lease or purchase new equipment.

The developers of the System/360 attacked that problem head-on. They conceived of a family of computers that would include machines of different sizes suitable for different applications, all of which would use the same instruction set and could share peripherals. To achieve this compatibility, they applied the principle of *modularity in design:* that is, the System/360's designers divided the designs of the processors and peripherals into *visible* and *hidden* information. IBM set up a Central Processor Control Office, which established and enforced the visible overall design rules that determined how the different modules of the machine would work together. The dozens of design teams scattered around the world had to adhere absolutely to these rules. But each team had full control over the hidden elements of design in its module—those elements that had no effect on other modules. (See "A Guide to Modularity.")

A Guide to Modularity

Modularity is a strategy for organizing complex products and processes efficiently. A *modular* system is composed of units (or modules) that are designed independently but still function as an integrated whole. Designers achieve modularity by partitioning information into *visible design rules* and *hidden design parameters*. Modularity is beneficial only if the partition is precise, unambiguous, and complete.

The visible design rules (also called *visible information*) are decisions that affect subsequent design decisions. Ideally, the visible design rules are established early in a design process and communicated broadly to those involved. Visible design rules fall into three categories:

- An *architecture,* which specifies what modules will be part of the system and what their functions will be.
- *Interfaces* that describe in detail how the modules will interact, including how they will fit together, connect, and communicate.
- *Standards* for testing a module's conformity to the design rules (can module X function in the system?) and for measuring one module's performance relative to another (how good is module X versus module Y?).

Practitioners sometimes lump all three elements of the visible information together and call them all simply "the architecture," "the interfaces," or "the standards."

The hidden design parameters (also called *hidden information*) are decisions that do not affect the design beyond the local module. Hidden elements can be chosen late and changed often and do not have to be communicated to anyone beyond the module design team.

When IBM employed this approach and also made the new systems compatible with existing software (by adding "emulator" modules), the result was a huge commercial and financial success for the company and its customers. Many of IBM's mainframe rivals were forced to abandon the market or seek niches focused on customers with highly specialized needs. But modularity also undermined IBM's dominance in the long run, as new companies produced their own so-called plug-compatible modules—printers, terminals, memory, software, and eventually even the central processing units themselves—that were compatible with, and could plug right into, the IBM machines. By following IBM's design rules but specializing in a particular area, an upstart company could often produce a module that was better than the ones IBM was making internally. Ultimately, the dynamic, innovative industry that has grown up around these modules developed entirely new kinds of computer systems that have taken away most of the mainframe's market share.

The fact that different companies (and different units of IBM) were working independently on modules enormously boosted the rate of innovation. By concentrating on a single module, each unit or company could push deeper into its workings. Having many companies focus on the design of a given module fostered numerous, parallel experiments. The module designers were free to try out a wide range of approaches as long as they obeyed the *design rules* ensuring that the modules would fit together. For an industry like computers, in which technological uncertainty is high and the best way to proceed is often unknown, the more experiments and the more flexibility each designer has to develop and test the experimental modules, the faster the industry is able to arrive at improved versions.

This freedom to experiment with product design is what distinguishes modular suppliers from ordinary subcontractors. For example, a team of disk drive designers has to obey the overall requirements of a personal computer, such as data transmission protocols, specifications for the size and shape of hardware, and standards for interfaces, to be sure that the module will function within the system as a whole. But otherwise, team members can design the disk drive in the way they think works best. The decisions they make need not be communicated to designers of other modules or even to the system's architects, the creators of the visible design rules. Rival disk-drive designers, by the same token, can experiment with completely different engineering approaches for their versions of the module as long as they, too, obey the visible design rules.[1]

Modularity Outside the Computer Industry

As a principle of production, modularity has a long history. Manufacturers have been using it for a century or more because it has always been easier to make complicated products by dividing the manufacturing process into modules or *cells*. Carmakers, for example, routinely manufacture the components of an automobile at different sites and then bring them together for final assembly. They can do so because they have precisely and completely specified the design of each part. In this context, the engineering design of a part (its dimensions and tolerances) serves as the visible information in the manufacturing system, allowing a complicated process to be split up among many factories and even outsourced to other suppliers. Those suppliers may experiment with production processes or logistics, but, unlike in the computer industry, they have historically had little or no input into the design of the components.

Modularity is comparatively rare not only in the actual design of products but also in their use. *Modularity in use* allows consumers to mix and match elements to come up with a final product that suits their tastes and needs. For example, to make a bed, consumers often buy bed frames, mattresses, pillows, linens, and covers from different manufacturers and even different retailers. They all fit together because the different manufacturers put out these goods according to standard sizes. Modularity in use can spur innovation in design: the manufacturers can independently experiment with new products and concepts, such as futon mattresses or fabric blends, and find ready consumer acceptance as long as their modules fit the standard dimensions.

If modularity brings so many advantages, why aren't all products (and processes) fully modular? It turns out that modular systems are much more difficult to design than comparable interconnected systems. The designers of modular systems must know a great deal about the inner workings of the overall product or process in order to develop the visible design rules necessary to make the modules function as a whole. They have to specify those rules in advance. And while designs at the modular level are proceeding independently, it may seem that all is going well; problems with incomplete or imperfect modularization tend to appear only when the modules come together and work poorly as an integrated whole.

IBM discovered that problem with the System/360, which took far more resources to develop than expected. In fact, had the developers initially realized the difficulties of ensuring modular integration, they might never have pursued the approach at all because they also underestimated the System/360's market value. Customers wanted it so much that their willingness to pay amply justified IBM's increased costs.

We have now entered a period of great advances in modularity. Breakthroughs in materials science and other fields have made it easier to obtain the deep product knowledge necessary to specify the design rules. For example, engineers now understand how metal reacts under force well enough to ensure modular coherence in body design and metal-forming processes for cars and big appliances. And improvements in computing, of course, have dramatically decreased the cost of capturing, processing, and storing that knowledge, reducing the cost of designing and testing different modules as well. Concurrent improvements in financial markets and innovative contractual arrangements are helping small companies find resources and form alliances to try out experiments and market new products or modules. In some industries, such as telecommunications and electric utilities, deregulation is freeing companies to divide the market along modular lines.

In automobile manufacturing, the big assemblers have been moving away from the tightly centralized design system that they have relied on for much of this century. Under intense pressure to reduce costs, accelerate the pace of innovation, and improve quality, automotive designers and engineers are now looking for ways to parcel out the design of their complex electromechanical system.

The first step has been to redefine the cells in the production processes. When managers at Mercedes-Benz planned their new sport-utility assembly plant in Alabama, for example, they realized that the complexities of the vehicle would require the plant to control a network of hundreds of suppliers according to an intricate schedule and to keep substantial inventory as a buffer against unexpected developments. Instead of trying to manage the supply system directly as a whole, they structured it into a smaller set of large production modules. The entire driver's cockpit, for example—including air bags, heating and air-conditioning systems, the instrument cluster, the steering column, and the wiring harness—is a separate module produced at a nearby plant owned by Delphi Automotive Systems, a unit of General Motors Corporation. Delphi is wholly responsible for producing the cockpit module according to certain specifications and scheduling requirements, so it can form its own network of dozens of suppliers for this module. Mercedes' specifications and the scheduling information become the visible information that module suppliers use to coordinate and control the network of parts suppliers and to build the modules required for final production.

Volkswagen has taken this approach even further in its new truck factory in Resende, Brazil. The company provides the factory where all modules are built and the trucks are assembled, but the independent suppliers obtain their own materials and hire their own workforces to build the separate modules. Volkswagen does not "make" the car, in the sense of producing or assembling it. But it does establish the architecture of the production process and the interfaces between cells, it sets the standards for quality that each supplier must meet, and it tests the modules and the trucks as they proceed from stage to stage.

So far, this shift in supplier responsibilities differs little from the numerous changes in supply-chain management that many industries are going through. By delegating the manufacturing process to many separate suppliers, each one of which adds value, the assembler gains flexibility and cuts costs. That amounts to a refinement of the pattern of modularity already established in production. Eventually, though, strategists at Mercedes and other automakers expect the newly strengthened module makers to take on most of the design responsibility as well—and that is the point at which modularity will pay off the most. As modularity becomes an established way of doing business, competition among module suppliers will intensify. Assemblers will look for the best-performing or lowest cost modules, spurring these increasingly sophisticated and independent suppliers into a race for innovation similar to the one already happening with computer modules. Computer-assisted design will facilitate this new wave of experimentation.

Some automotive suppliers are already moving in that direction by consolidating their industry around particular modules. Lear Seating Corporation, Magna International, and Johnson Controls have been buying related suppliers, each attempting to become the worldwide leader in the production of entire car interiors. The big car manufacturers are indirectly encouraging this process by asking their suppliers to participate in the design of modules. Indeed, GM recently gave Magna total responsibility for overseeing development for the interior of the next-generation Cadillac Catera.

In addition to products, a wide range of services are also being modularized—most notably in the financial services industry, where the process is far along. Nothing is easier to modularize than stocks and other securities. Financial services are purely intangible, having no hard surfaces, no difficult shapes, no electrical pins or wires, and no complex computer code. Because the science of finance is sophisticated and highly developed, these services are relatively easy to define, analyze, and split apart. The design rules for financial transactions arise from centuries-old traditions of bookkeeping combined with modern legal and industry standards and the conventions of the securities exchanges.

As a result, providers need not take responsibility for all aspects of delivering their financial services. The tasks of managing a portfolio of securities, for example—selecting assets, conducting trades, keeping records, transferring ownership, reporting status and sending out statements, and performing custody services—can be readily broken apart and seamlessly performed by separate suppliers. Some major institutions have opted to specialize in one such area: Boston's State Street Bank in custody services, for example.

Other institutions, while modularizing their products, still seek to own and control those modules, as IBM tried to control the System/360. For example, Fidelity, the big, mass-market provider of money management services, has traditionally kept most aspects of its operations in-house. However, under pressure to reduce costs, it recently broke with that practice, announcing that Bankers Trust Company would manage $11 billion worth of stock index funds on its behalf. Index funds are a low-margin business whose performance is easily measured. Under the new arrangement, Bankers Trust's index-fund management services have become a hidden module in Fidelity's overall portfolio offerings, much as Volkswagen's suppliers operate as hidden modules in the Resende factory system.

The other result of the intrinsic modularity of financial instruments has been an enormous boost in innovation. By combining advanced scientific methods with high-speed computers, for example, designers can split up securities into smaller units that can then be reconfigured into derivative financial products. Such innovations have made global financial markets more fluid so that capital now flows easily even between countries with very different financial practices.

Competing in a Modular Environment

Modularity does more than accelerate the pace of change or heighten competitive pressures. It also transforms relations among companies. Module designers rapidly move in and out of joint ventures, technology alliances, subcontracts, employment agreements, and financial arrangements as they compete in a relentless race to innovate. In such markets, revenue and profits are far more dispersed than they would be in traditional industries. Even such companies as Intel and Microsoft, which have substantial market power by virtue of their control over key subsets of visible information, account for less of the total market value of all computer companies than industry leaders typically do.

Being part of a shifting modular cluster of hundreds of companies in a constantly innovating industry is different from being one of a few dominant companies in a stable industry. No strategy or sequence of moves will always work; as in chess, a good move depends on the layout of the board, the pieces one controls, and the skill and resources of one's opponent. Nevertheless, the dual structure of a modular marketplace requires managers to choose carefully from two main strategies. A company can compete as an architect, creating the visible information, or design rules, for a product made up of modules. Or it can compete as a designer of modules that conform to the architecture, interfaces, and test protocols of others. Both strategies require companies to understand products at a deep level and be able to predict how modules will evolve, but they differ in a number of important ways.

For an architect, advantage comes from attracting module designers to its design rules by convincing them that this architecture will prevail in the marketplace. For the module maker, advantage comes from mastering the hidden information of the design and from superior execution in bringing its module to market. As opportunities emerge, the module maker must move quickly to fill a need and then move elsewhere or reach new levels of performance as the market becomes crowded.

Following the example of Intel and Microsoft, it is tempting to say that companies should aim to control the visible design rules by developing proprietary architectures and leave the mundane details of hidden modules to others. And it is true that the position of architect is powerful and can be very profitable. But a challenger can rely on modularity to mix and match its own capabilities with those of others and do an end-run around an architect.

That is what happened in the workstation market in the 1980s. Both of the leading companies, Apollo Computer and Sun Microsystems, relied heavily on other companies for the design and production of most of the modules that formed their workstations. But Apollo's founders, who emphasized high performance in their product, designed a proprietary architecture based on their own operating and network management systems. Although some modules, such as the microprocessor, were bought off the shelf, much of the hardware was designed in-house. The various parts of the design were highly interdependent, which Apollo's designers believed was necessary to achieve high levels of performance in the final product.

Sun's founders, by contrast, emphasized low costs and rapid time to market. They relied on a simplified, nonproprietary architecture built with off-the-shelf hardware and software, including the widely available UNIX operating system. Because its module makers did not have to design special modules to fit into its system, Sun was free of the investments in software and hardware design Apollo required and could bring products to market quickly while keeping capital costs low. To make up for the performance penalty incurred by using generic modules, Sun developed two proprietary, hidden hardware modules to link the microprocessor efficiently to the workstation's internal memory.

In terms of sheer performance, observers judged Apollo's workstation to be slightly better, but Sun had the cost advantage. Sun's reliance on other module makers proved superior in other respects as well. Many end users relied on the UNIX operating system in other networks or applications and preferred a workstation that ran on UNIX rather than one that used a more proprietary operating system. Taking advantage of its edge in capital productivity, Sun opted for an aggressive strategy of rapid growth and product improvement.

Soon, Apollo found itself short of capital and its products' performance fell further and further behind Sun's. The flexibility and leanness Sun gained through its nonproprietary approach overcame the performance advantages Apollo had been enjoying through its proprietary strategy. Sun could offer customers an excellent product at an attractive price, earn superb margins, and employ much less capital in the process.

However, Sun's design gave it no enduring competitive edge. Because Sun controlled only the two hidden modules in the workstation, it could not lock its customers into its own proprietary operating system or network protocols. Sun did develop original ideas about how to combine existing modules into an effective system, but any competitor could do the same since the architecture—the visible information behind the workstation design—was easy to copy and could not be patented.

Indeed, minicomputer makers saw that workstations would threaten their business and engineering markets, and they soon offered rival products, while personal computer makers (whose designs were already extremely modular) saw an opportunity to move into a higher-margin niche. To protect itself, Sun shifted gears and sought greater control over the visible information in its own system. Sun hoped to use equity financing from AT&T, which controlled UNIX, to gain a favored role in designing future versions of the operating system. If Sun could control the evolution of UNIX, it could bring the next generation of workstations to market faster than its rivals could. But the minicomputer makers, which licensed UNIX for their existing systems, immediately saw the threat posed by the Sun-AT&T alliance, and they forced AT&T to back away from Sun. The workstation market remained wide open, and when Sun stumbled in bringing out a new generation of workstations, rivals gained ground with their own offerings. The race was on—and it continues.

Needed: Knowledgeable Leaders

Because modularity boosts the rate of innovation, it shrinks the time business leaders have to respond to competitors' moves. We may laugh about the concept of an "Internet year," but it's no joke. As more and more industries pursue modularity, their general managers, like those in the computer industry, will have to cope with higher rates of innovation and swifter change.

As a rule, managers will have to become much more attuned to all sorts of developments in the design of products, both inside and outside their own companies. It won't be enough to know what their direct competitors are doing—innovations in other modules and in the overall product architecture, as well as shifting alliances elsewhere in the industry, may spell trouble or present opportunities. Success in the marketplace will depend on mapping a much larger competitive terrain and linking one's own capabilities and options with those emerging elsewhere, possibly in companies very different from one's own.

Those capabilities and options involve not only product technologies but also financial resources and the skills of employees. Managers engaged with modular design efforts must be adept at forging new financial relationships and employment contracts, and they must enter into innovative technology ventures and alliances. Harvard Business School professor Howard Stevenson has described entrepreneurship as "the pursuit of opportunity beyond the resources currently controlled," and that's a good framework for thinking about modular leadership at even the biggest companies. (See "How Palm Computing Became an Architect" and "How Quantum Mines Hidden Knowledge.")

How Palm Computing Became an Architect

In 1992, Jeff Hawkins founded Palm Computing to develop and market a handheld computing device for the consumer market. Having already created the basic software for handwriting recognition, he intended to concentrate on refining that software and developing related applications for this new market. His plan was to rely on partners for the basic architecture, hardware, operating system software, and marketing. Venture capitalists funded Palm's own development. The handwriting recognition software became the key hidden module around which a consortium of companies formed to produce the complete product.

Sales of the first generation of products from both the consortium and its rivals, however, were poor, and Palm's partners had little interest in pursuing the next generation. Convinced that capitalizing on Palm's ability to connect the device directly to a PC would unlock the potential for sales, Hawkins and his chief executive, Donna Dubinsky, decided to shift course. If they couldn't get partners to develop the new concept, they would handle it themselves— at least the visible parts, which included the device's interface protocols and its operating system. Palm would have to become an architect, taking control of both the visible information and the hidden information in the handwriting recognition module. But to do so, Hawkins and Dubinsky needed a partner with deeper pockets than any venture capital firm would provide.

None of the companies in Palm's previous consortium was willing to help. Palm spread its net as far as US Robotics, the largest maker of modems. US Robotics was so taken with the concept for and development of Palm's product that it bought the company. With that backing, Palm was able to take the product into full production and get the marketing muscle it needed. The result was the Pilot, or what Palm calls a Personal Connected Organizer, which has been a tremendous success in the marketplace. Palm remains in control of the operating system and the handwriting recognition software in the Pilot but relies on other designers for hardware and for links to software that runs on PCs.

Palm's strategy with the Pilot worked as Hawkins and Dubinsky had intended. In order for its architecture to be accepted by customers and outside developers, Palm had to create a compelling concept that other module makers would accept, with attractive features and pricing, and bring the device to market quickly. Hawkins's initial strategy—to be a hidden-module producer while partners delivered the architecture—might have worked with a more familiar product, but the handheld-computer market was too unformed for it to work in that context. So, when the other members of the consortium balked in the second round of the design process, Palm had to take the lead role in developing both the proof of concept and a complete set of accessible design rules for the system as a whole.

We are grateful to Myra Hart for sharing with us her ongoing research on Palm. She describes the company in detail in her cases "Palm Computing, Inc. (A)," HBS case no. 396245, and "Palm Computing, Inc. 1995: Financing Challenges," HBS case no. 898090.

How Quantum Mines Hidden Knowledge

Quantum Corporation began in 1980 as a maker of 8-inch disk storage drives for the minicomputer market. After the company fell behind as the industry shifted to 5.25-inch drives, a team led by Stephen M. Berkley and Dave Brown rescued it with an aggressive strategy, applying their storage expertise to developing a 3.5-inch add-on drive for the personal computer market. The product worked, but competing in this sector required higher volumes and tighter tolerances than Quantum was used to. Instead of trying to meet those demands internally, Berkley and Brown decided to keep the company focused on technology and to form an alliance with Matsushita-Kotobuki Electronics Industries (MKE), a division of the Matsushita Group, to handle the high-volume, high-precision manufacturing. With the new alliance in place, Quantum and MKE worked to develop tightly integrated design capabilities that spanned the two companies. The products resulting from those processes allowed Quantum to compete successfully in the market for drives installed as original equipment in personal computers.

Quantum has maintained a high rate of product innovation by exploiting modularity in the design of its own products and in its own organization.

Separate, small teams work on the design and the production of each submodule, and the company's leaders have developed an unusually clear operating framework within which to coordinate the efforts of the teams while still freeing them to innovate effectively.

In addition to focusing on technology, the company has survived in the intensely competitive disk-drive industry by paying close attention to the companies that assemble personal computers. Quantum has become the preferred supplier for many of the assemblers because its careful attention to developments in the visible information for disk drives has enabled its drives to fit seamlessly into the assemblers' systems. Quantum's general managers have a deep reservoir of knowledge about both storage technology and the players in the sector, which helps them map the landscape, anticipating which segments of the computer market are set to go into decline and where emerging opportunities will arise. Early on, they saw the implications of the Internet and corporate intranets, and with help from a timely purchase of Digital Equipment Corporation's stagnating storage business, they had a head start in meeting the voracious demand for storage capacity that has been created by burgeoning networks. Despite what some observers might see as a weak position (because the company must depend on the visible information that other companies give out) Quantum has prospered, recently reporting strong profits and gains in stock price.

We are grateful to Steven Wheelwright and Clayton Christensen for sharing with us their ongoing research on Quantum. They describe the company in more detail in their case "Quantum Corp.: Business and Product Teams," HBS case no. 692023.

At the same time that modularity boosts the rate of innovation, it also heightens the degree of uncertainty in the design process. There is no way for managers to know which of many experimental approaches will win out in the marketplace. To prepare for sudden and dramatic changes in markets, therefore, managers need to be able to choose from an often complex array of technologies, skills, and financial options. Creating, watching, and nurturing a portfolio of such options will become more important than the pursuit of static efficiency per se.

To compete in a world of modularity, leaders must also redesign their internal organizations. In order to create superior modules, they need the flexibility to move quickly to market and make use of rapidly changing technologies, but they must also ensure that the modules conform to the architecture. The answer to this dilemma is modularity within the organization. Just as modularity in design boosts innovation in products by freeing designers to experiment, so managers can speed up development cycles for individual modules by splitting the work among independent teams, each pursuing a different submodule or different path to improvement.

Employing a modular approach to design complicates the task of managers who want to stabilize the manufacturing process or control inventories because it expands the range of possible product varieties. But the approach also allows engineers to create families of parts that share common characteristics and thus can all be made in the same way, using, for example, changes in machine settings that are well understood. Moreover, the growing power of information technology is giving managers more precise and timely information about sales and distribution channels, thus enhancing the efficiency of a modular production system.

For those organizational processes to succeed, however, the output of the various decentralized teams (including the designers at partner companies) must be tightly integrated. As with a product, the key to integration in the organization is the visible information. This is where leadership is critical. Despite what many observers of leadership are now saying, the heads of these companies must do more than provide a vision or goals for the decentralized development teams. They must also delineate and communicate a detailed operating framework within which each of the teams must work.

Such a framework begins by articulating the strategy and plans for the product line's evolution into which the work of the development teams needs to fit over time. But the framework also has to extend into the work of the teams themselves. It must, for example, establish principles for matching appropriate types of teams to each type of project. It must specify the size of the teams and make clear what roles senior management, the core design team, and support groups should play in carrying out the project's work. Finally, the framework must define processes by which progress will be measured and products released to the market. The framework may also address values that should guide the teams in their work (such as leading by example). Like the visible information in a modular product, this organizational framework establishes an overall structure within which teams can operate, provides ways for different teams and other groups to interact, and defines standards for testing the merit of the teams' work.

Without careful direction, the teams would find it easy to pursue initiatives that may have individual merit but stray from the company's defining concepts.

Just like a modular product that lacks good interfaces between modules, an organization built around decentralized teams that fail to function according to a clear and effective framework will suffer from miscues and delays. Fast changing and dynamic markets—like those for computers—are unforgiving. The well-publicized problems of many computer companies have often been rooted in inadequate coordination of their development teams as they created new products. Less obvious, but equally important, are the problems that arise when teams fail to communicate the hidden information—the knowledge they develop about module technology—with the rest of the organization. That lack of communication, we have found, causes organizations to commit the same costly mistakes over and over again.

To take full advantage of modularity, companies need highly skilled, independent-minded employees eager to innovate. These designers and engineers do not respond to tight controls; many reject traditional forms of management and will seek employment elsewhere rather than submit to them. Such employees do, however, respond to informed leadership—to managers who can make reasoned arguments that will persuade employees to hold fast to the central operating framework. Managers must learn how to allow members of the organization the independence to probe and experiment while directing them to stay on the right overall course. The best analogy may be in biology, where complex organisms have been able to evolve into an astonishing variety of forms only by obeying immutable rules of development.

A century ago, the railroads showed managers how to control enormous organizations and masses of capital. In the world fashioned by computers, managers will control less and will need to know more. As modularity drives the evolution of much of the economy, general managers' greatest challenge will be to gain an intimate understanding of the knowledge behind their products. Technology can't be a black box to them because their ability to position the company, respond to market changes, and guide internal innovation depends on this knowledge. Leaders cannot manage knowledge at a distance merely by hiring knowledgeable people and giving them adequate resources. They need to be closely involved in shaping and directing the way knowledge is created and used. Details about the inner workings of products may seem to be merely technical engineering matters, but in the context of intense competition and fast changing technology, the success of whole strategies may hinge on such seemingly minor details.

Further Reading

For more information on modular product design, see Steven D. Eppinger, Daniel E. Whitney, Robert P. Smith, and David Gebala, "A Model-Based Method for Organizing Tasks in Product Development," *Research in Engineering Design* 6, 1991. For more about modular processes, see James L. Nevins and Daniel E. Whitney, *Concurrent Design of Products and Processes* (New York: McGraw-Hill, 1989). For more information on the design of financial securities and the global financial system, see Robert C. Merton and Zvi Bodie, "A Conceptual Framework for Analyzing the Financial Environment" and "Financial Infrastructure and Public Policy: A Functional Perspective," in *The Global Financial System: A Functional Perspective*, (Boston, Massachusetts: Harvard Business School Press, 1995).

For descriptions of how companies compete in industries using modular products, see Richard N. Langlois and Paul L. Robertson, "Networks and Innovation in a Modular System: Lessons from the Microcomputer and Stereo Component Industries," *Research Policy*, August 1992; Charles R. Morris and Charles H. Ferguson, "How Architecture Wins Technology Wars," HBR March–April 1993; Raghu Garud and Arun Kumaraswamy, "Changing Competitive Dynamics in Network Industries: An Exploration of Sun Microsystems' Open Systems Strategy," *Strategic Management Journal*, July 1993, p. 351; and Clayton M. Christensen and Richard S. Rosenbloom, "Explaining the Attacker's Advantage: Technological Paradigms, Organizational Dynamics, and the Value Network," *Research Policy*, March 1995, p. 233.

Note

1. Practical knowledge of modularity has come largely from the computer industry. The term *architecture* was first used in connection with computers by the designers of the System/360: Gene M. Amdahl, Gerrit A. Blaauw, and Frederick P. Brooks, Jr., in "Architecture of the

IBM System/360," *IBM Journal of Research and Development*, April 1964, p. 86. The scientific field of computer architecture was established by C. Gordon Bell and Allen Newell in *Computer Structures: Readings and Examples* (New York: McGraw-Hill, 1971). The principle of *information hiding* was first put forward in 1972 by David L. Parnas in "A Technique for Software Module Specification with Examples," *Communications of the ACM*, May 1972, p. 330. The term *design rules* was first used by Carver Mead and Lynn Conway in *Introduction to VLSI Systems* (Reading, Massachusetts: Addison-Wesley, 1980). Sun's architectural innovations, described in the text, were based on the work of John L. Hennessy and David A. Patterson, later summarized in their text *Computer Architecture: A Quantitative Approach* (San Mateo, California: Morgan Kaufman Publishers, 1990).

4
Do You Want to Keep Your Customers Forever?

B. Joseph Pine II, Don Peppers, and Martha Rogers

Customers, whether consumers or businesses, do not want more choices. They want exactly what they want—when, where, and how they want it—and technology now makes it possible for companies to give it to them. Interactive and database technology permits companies to amass huge amounts of data on individual customers' needs and preferences. And information technology and flexible manufacturing systems enable companies to customize large volumes of goods or services for individual customers at a relatively low cost. But few companies are exploiting this potential. Most managers continue to view the world through the twin lenses of mass marketing and mass production. To handle their increasingly turbulent and fragmented markets, they try to churn out a much greater variety of goods and services and to target ever finer market segments with more tailored advertising messages. But these managers only end up bombarding their customers with too many choices.

A company that aspires to give customers exactly what they want must look at the world through new lenses. It must use technology to become two things: a *mass customizer* that efficiently provides individually customized goods and services, and a *one-to-one marketer* that elicits information from each customer about his or her specific needs and preferences. The twin logic of mass customization and one-to-one marketing binds producer and consumer together in what we call a *learning relationship*—an ongoing connection that becomes smarter as the two interact with each other, collaborating to meet the consumer's needs over time.

In learning relationships, individual customers teach the company more and more about their preferences and needs, giving the company an immense competitive advantage. The more customers teach the company, the better it becomes at providing exactly what they want—exactly how they want it—and the more difficult it will be for a competitor to entice them away. Even if a competitor were to build the exact same capabilities, a customer already involved in a learning relationship with the company would have to spend an inordinate amount of time and energy teaching the competitor what the company already knows.

Because of this singularly powerful competitive advantage, a company that can cultivate learning relationships with its customers should be able to retain their business virtually forever—provided that it continues to supply high-quality customized products or services at reasonably competitive prices and does not miss the next technology wave. (Learning relationships would not have saved a buggy-whip manufacturer from the automobile.)

One company that excels at building learning relationships with its customers is named, appropriately enough, Individual, Inc. This Burlington, Massachusetts, company, which competes with wire, clipping, and information-retrieval services, provides published news stories selected to fit the specific, ever changing interests of each client. Instead of having to sort through a mountain of clippings or having to master the arcane commands needed to search databases, Individual's customers—which include such diverse companies as MCI Telecommunications, McKinsey & Company, Avon Products, and Fidelity Investments—effortlessly receive timely, fresh, relevant articles delivered right to their desks by fax, groupware (such as Lotus Notes), on-line computer services, the Internet, or electronic mail.

When someone signs up for Individual's *First!* service, the company assigns an editorial manager to determine what sort of information the client wants. The editorial manager and the client reduce those requests to simple descriptions, such as articles about new uses of information technology in home health care or about new products developed by Japanese semiconductor companies. The editorial manager enters the requests into Individual's SMART software system (for System for Manipulation and Retrieval of Text). Then SMART takes over. Every business day, the system searches 400 sources containing more than 12,000 articles for those pieces that will most likely fit the client's needs, and it delivers them by whatever method the client has chosen.

Every week, Individual asks a new client (by fax or computer) to rate each article as "not relevant," "somewhat relevant," or "very relevant." The responses are fed into the system, making SMART even smarter. In the first week of service, most customers find only 40% to 60% of the articles to be somewhat or very relevant. By the fourth or fifth week, SMART has increased those ratings to a targeted 80% to 90%. Once it has achieved that level, Individual reduces the frequency of the ratings to once a month, which still enables it to keep abreast of customers' changing needs.

Individual also responds constantly to clients' requests for new sources and ways of receiving information. Sun Microsystems, for example, asked the company to place *First!* on its internal Internet server. Once Individual provided this service, it discovered that many other clients that also depended on the Internet for sending and sharing information wanted to receive the service in the same way. Such responsiveness is undoubtedly one reason why Individual, which has more than 30,000 users and more than 4,000 accounts, enjoys a customer-retention rate of 85% to 90%. But there is also another reason: because of the time and energy each client expends in teaching the company which articles are relevant and which are not, switching to a competitor would require the client to make that investment all over again.

From Mass Production to Mass Customization

Although Individual uses information and interactive technology to its fullest, most managers fail to understand that variety is not the same thing as customization. Customization means manufacturing a product or delivering a service *in response* to a particular customer's needs, and mass customization means doing it in a cost-effective way. Mass customization calls for a customer-centered orientation in production and delivery processes, requiring the company to collaborate with individual customers to design each one's desired product or service, which is then constructed from a base of pre-engineered modules that can be assembled in myriad ways.

In contrast, product-centered mass production and mass marketing call for pushing options (and inventory) into distribution channels and hoping that each new option is embraced by enough customers to make its production worthwhile. It requires customers to hunt for the single product or service they want from among an ever growing array of alternatives.

Consider grocery stores. According to *New Products News,* the number of new products, including line extensions, introduced in grocery stores each year increased from less than 3,000 in 1980 to more than 10,000 in 1988 and more than 17,000 in 1993. And *Progressive Grocer* reports that the number of stock-keeping units in the average supermarket doubled to more than 30,000 between 1980 and 1994. The same trend can be seen in many service industries: witness the proliferation of affinity credit cards and the numerous options offered by telephone companies.

Companies are also deluging consumers with a wider variety of messages. And, of course, there is a greater array of media for carrying them: direct mail, telemarketing, special newspaper supplements, and a larger number of television channels, among others.

For example, the average newspaper weighs 55% more today than it did just ten years ago, mainly because of supplements designed to carry specially targeted advertising. The problem with such supplements is that they are distributed to every subscriber. Nongardeners still receive the gardening supplement, and people reading the paper before heading to the office still get the work-at-home supplement. So the supplements really aren't so special after all.

Mass marketers use information technology to define the most likely customers for the products they want to sell. For the most part, the information comes from simple transactional records (such as customer purchases and invoices) and public information (such as vehicle registrations, address-change forms, and census data) compiled by companies like R.L. Polk and Donnelley Marketing. From those data, the mass marketer generates a list of the most likely prospects and solicits them with offers or messages that the marketer has attempted to customize by guessing their tastes. By contrast, the one-to-one marketer conducts a dialogue with each customer—one at a time—and uses the increasingly more detailed feedback to find the best products or services for that customer. Although many companies are moving toward this model, few have fully implemented it yet or combined it with mass customization.

Take Hallmark Cards and American Greetings, the leaders of the variety-intensive greeting card industry. Both companies have installed electronic kiosks in stores and other public places to enable people to create their own greeting cards. Consumers can touch the screen of either company's kiosk, quickly select the type of card they need (for example, anniversary or birthday card), browse through a number of selections, and then modify them or compose their own wording to express exactly the right sentiment. The card is printed in a minute or so.

Both companies seem pleased with the performance of their mass customization businesses, but neither has fully exploited its potential. The graphics for the cards are all preset (so only the wording can be customized), and there is little organization (so browsing through the choices can be time consuming). The greatest weakness of the electronic kiosks, however, is the absence of a system for recording individual customers' preferences. Each time someone uses the system, he or she must start all over again.

If a greeting card company were to harness the full power of mass customization and one-to-one marketing, it would be able to remember the important occasions in your life and remind you to buy a card. It would make suggestions based on your past purchases. Its kiosk would display past selections, either to ensure that you don't commit the faux pas of sending the same card to the same person twice or to give you the option of sending the same funny card to another person—appropriately personalized, of course. Perhaps the company would mail your cards or ship them across the Internet for you so they would arrive at the appointed time. Maybe the company would be able to remind you to send a card, allow you to design it, and arrange for its delivery on your personal computer through an on-line service that would let you incorporate your own graphics or photographs. It might even find your design so good that it would ask your permission to add it to its inventory.

Certainly, not every customer would want to invest the time that such a relationship would require. Neither would every customer buy enough cards to make such a relationship worthwhile for the company. But the advantages to a greeting card company of establishing and cultivating a learning relationship with customers who buy cards frequently are immense. Because every card sold to those customers will be tailored precisely to their needs, the company will be able to charge them a premium and its profit margins will increase. And because the company will be equipped to ensure that the customer never forgets an occasion, it will sell more cards to that customer. The company's product development will become more effective because of the expanded ability to understand exactly who is buying what, when, and why—not to mention the ability to use new ideas that customers could provide.

But, most important of all, the company will retain more customers, especially the most valuable ones: frequent purchasers. The more customers teach the company about their individual tastes, celebration occasions, and card recipients (addresses, relationships, and so forth), the more reluctant they will be to repeat that process with another supplier. As long as the company fulfills its end of the bargain, a competitor should never be able to entice away its customers. The battle will be limited to attracting new ones.

When Are Learning Relationships Appropriate?

As compelling and powerful as the benefits of learning relationships are, this radically different business model cannot be applied in the same way by everyone. Companies such as home builders, real estate brokers, and appliance manufacturers—which do not interact frequently with end users—cannot learn enough to make a learning relationship with those customers work. But they might find it beneficial to develop such relationships with general contractors. Similarly, makers of products like paper clips, whose revenue or profit margin per customer is too low to justify building individual learning relationships with customers, might find it advantageous to cultivate learning relationships with office-supply chains, which interact directly with end users.

Even producers of commodities such as wheat or natural gas, which cannot be customized easily, and of commodity-like products bought mainly on the basis of price have much to gain from this approach. Learning relationships can enable such companies to design services that differentiate their offerings. This is the strategy that Bandag, which sells truck-tire retreads to more than 500 dealer-installers around the country, is pursuing.

Bandag's retreads are essentially a commodity because they are comparable in price and quality to those of competitors. To break out of the pack, Bandag is providing additional services. For example, it assists its dealers in filing and collecting on warranty claims from tire manufacturers and will soon begin offering comprehensive fleet-management services to its largest national accounts.

Bandag plans to embed computer chips in the rubber of newly retreaded tires to gauge each tire's pressure and temperature and to count its revolutions. That information will enable the company not only to tell each customer the optimal time to retread each tire (thus reducing downtime caused by blowouts) but also to help it improve its fleet's operations.

Because of the current high cost of building such capabilities, many manufacturers, service providers, and retailers may find, as Bandag did, that it pays to establish learning relationships only with their best customers. But as advances in information technology continue to drive down the cost of building learning relationships, they will make economic sense in many more businesses and for a wider spectrum of customers. Many types of industries are already ripe for revolution. They include:

Complex Products or Services. Most people do not want to work their way through hundreds or thousands of options, features, pricing structures, delivery methods, and networks to figure out which product or service is best for them. One solution is for companies to collaborate with customers in custom-designing the product, as Andersen Corporation, the window manufacturer based in Bayport, Minnesota, is doing. It resolved the information-overload problem for its customers (individual home owners and building contractors) by developing a multimedia system called the Window of Knowledge. A sales representative uses a workstation that features 50,000 possible window components to help customers design their own windows. The system automatically generates error-free quotations and manufacturing specifications, which can be saved for future use. The resulting database of window configurations deepens Andersen's understanding of how its business is performing.

Big-Ticket Items. A company that succeeds in customizing all aspects of owning an expensive product or using a premium service stands to gain a competitive advantage over its rivals. Consider automobiles. A car buyer, over his or her lifetime, can generate hundreds of thousands of dollars' worth of business when financing, service, and referrals, as well as the original purchase, are taken into account. All together, they represent an enormous opportunity for companies that cater to customers' individual preferences. The same opportunities apply to big-ticket commercial offerings, including machinery, information systems, outsourcing, and consulting. (See "How to Gain Customers Forever.")

How to Gain Customers Forever

Industrial companies that sell to other businesses can benefit just as much from learning relationships as companies that sell products or services to consumers. Consider the case of Ross Controls (formerly the Ross Operating Valve Company) of Troy, Michigan, a 70-year-old manufacturer of pneumatic valves and air-control systems. Through what it calls the ROSS/FLEX process, Ross learns about its customers' needs, collaborates with them to come up with designs precisely tailored to help them meet those needs, and quickly and efficiently makes the customized products. The process has enabled the medium-size manufacturer to forge learning relationships with such companies as General Motors, Knight Industries, Reynolds Aluminum, and Japan's Yamamura Glass.

For example, Ross is currently supplying GM's Metal Fabricating Division with 600 integrated-valve systems. Based on a common platform but individually customized for a particular stamping press, each integrated system performs better than the valves it is replacing at one-third the price.

Two elements have enabled Ross to transform itself from a sleepy industrial manufacturer into a dynamic organization that cultivates learning relationships with its customers:

A Desire to Listen to and Collaborate with Each Customer. This involves spending time on the phone, faxing ideas back and forth, and often visiting plants to see how pneumatic systems are to be used in the customer's manufacturing process. And once a system is designed to solve the customer's problem, Ross gets feedback from prototypes and encourages the customer to make continuous upgrades to its valve designs, yielding more precisely tailored designs over time. Ross then stores them in a library of design platforms, components, and computer instructions for its manufacturing equipment so it does not have to start from scratch every time it works with a customer on a new project.

The Capability to Turn Complex Designs into Products. Through the effective use of computer-aided design (CAD) and computer numerically controlled (CNC) machines, Ross can electronically transmit tooling instructions directly from engineering workstations to multimillion-dollar production equipment, which can turn around new designs in as little as a day. But obviously, computer-aided-design and manufacturing equipment alone does not enable a company to mass-customize. Information about each customer's needs is also essential. To obtain such information, Ross created a crew of "integrators," each of whom is assigned to a given customer. The integrator talks with the customer, produces the valve designs, and determines the manufacturing specifications, including the instructions for the CNC machines. Using the CAD system, the integrator draws from the library's contents whenever possible to create a customized design and the computer coding required to make the product.

The ROSS/FLEX process has helped Ross boost the custom portion of its business from 5% to 20% of its revenues in the past four years. But the company is not yet satisfied with its ability to build learning relationships. It intends to add an interactive audio and video communications setup that will include a "what you see is what I see" CAD system so that an integrator and a customer do not have to be in the same place to collaborate on a design. And it plans to automate the access to its library so that integrators and customers—even on their own—can generate a wider range of designs and execute each one more quickly.

When Ross started down this road eight years ago, its primary goal was to gain customers for life by expanding the company's capabilities to meet each one's changing needs. It is clearly making a lot of progress. At a time when GM is reexamining virtually all its supplier relations, its Metal Fabricating Division won't go to any company but Ross for pneumatic valves and won't let its suppliers, either. Knight Industries, a supplier of ergonomic material-handling equipment, gives Ross 100% of its custom business and about 70% of its standard (catalog) business. When a competitor tried to woo Knight away, its president, James Zaguroli, Jr., responded, "Why would I switch to you? You're already five product generations behind where I am with Ross."

Digitizable Products and Services. Anything that can be digitized can be customized. If such products are purchased frequently, providing a discernible pattern of personal preferences, they may be ideally suited for one-to-one marketing as well. Obvious candidates include not only greeting cards but also software, periodicals, telecommunication services, and entertainment products such as movies, videos, games, and recorded music. Indeed, many companies in these businesses are working to develop learning relationships.

On-line Services. Providers of on-line services already offer a broad spectrum of choices—including electronic shopping, special-interest forums, entertainment, news, and financial services—but few offer tailored convenience. Currently, the user must navigate through choice after choice. A competitor that learns a customer's wants and needs could navigate cyberspace on behalf of that customer and cull only the relevant choices.

Luxury and Specialty Products. Many businesses (such as apparel, perfume and cosmetics, athletic equipment, and fine wine) have customers with complex individual tastes. For example, people differ not only in their physical measurements but also in how they prefer their clothes to fit and look. Levi Strauss is capitalizing on these differences by mass-customizing blue jeans for women, using technology supplied by Custom Clothing Technology Corporation of Newton, Massachusetts. After a customer has her measurements taken in a store, she tries on a pair or two of jeans to determine her exact preference. The information is then sent to the factory for prompt production. Although Levi Strauss is currently limiting the program to one style of jeans, the approach offers the company tremendous opportunities for building learning relationships.

Retailing Services. In many industries, retailers have a big advantage over manufacturers in building learning relationships with end users, especially when customers want to touch, feel, and browse (clothing, shoes, and books) or when the product is immediately consumed (for example, in restaurants and bars). They also have the edge when individual customers do not buy a large amount of any one manufacturer's products (such as groceries and packaged goods). That is because the retailer is in a better position to see patterns in a customer's purchases and because it might be more expensive for the manufacturer to build learning relationships. Finally, many retailers offer consumers not products per se but service, and services can be mass-customized more readily than most products can. (See "How Peapod Is Customizing the Virtual Supermarket.")

How Peapod Is Customizing the Virtual Supermarket

One company that is exploiting learning relationships in retailing services is Peapod, a grocery-shopping and delivery service based in Evanston, Illinois. Its customers—currently in Chicago and San Francisco—buy a software application for $29.95 that enables them to access Peapod's database through an on-line computer service. They then pay $4.95 per month for the service and a per-order charge of $5 plus 5% of the order amount. Peapod's back office is linked into the mainframe databases of the supermarkets at which it shops for its customers (Jewel in Chicago and Safeway in San Francisco), allowing it to provide all the supermarkets' stock-keeping units and shelf prices electronically to its customers.

Rather than automating the trip to a retail store, as other on-line providers are doing, Peapod is using interactive technology to change the shopping experience altogether. It lets each customer create the virtual supermarket that best suits him or her. Using a personal computer, customers can shop in the way they prefer. They can request a list of items by category (snack foods), by item (potato chips), by brand (Frito-Lay), or even by what is on sale in the store on a given day. Within categories, they can choose to have the items arranged alphabetically by brand, by package size, by unit price, or even by nutritional value. Customers also can create and save for repeated use standard and special shopping lists (baby items, barbecue needs, and the like).

Peapod teaches its customers to shop so effectively in its virtual supermarket that most of them discover that—despite the company's rates—they *save* money because they use more coupons, do better comparison shopping, and buy fewer impulse items than they would if they shopped at a real supermarket. In addition, they save time and have more control over it because they can shop from home or work whenever they want.

Peapod has found that every interaction with a customer is an opportunity to learn. At the end of each shopping session, it asks the customer, "How did we do on the last order?" Peapod gets feedback on 35% of orders; most companies consider a 10% response rate to customer-satisfaction surveys to be good. And more than 80% of Peapod's customers have responded at one time or another. The feedback has prompted the company to institute a variety of changes and options, including providing nutritional information, making deliveries within a half-hour window (for an additional $4.95) rather than the usual 90-minute window, accepting detailed requests (such as three ripe and three unripe tomatoes), and delivering alcoholic beverages.

Peapod views delivery as another opportunity to learn about customers' preferences. It asks its deliverers to find out where customers would like the groceries left when they're not at home and anything else that will enhance the relationship. They fill out an "interaction record" for every delivery to track those preferences (as well as entering basic service metrics, such as the time of the delivery).

Even with the rates it charges, Peapod has to be efficient and effective to make money in what is a low-margin business. That is why it mass-customizes all shopping and delivery processes. Each order is filled by a generalist, who shops the aisles of the store, and as-needed specialists, who provide the produce, meats, deli, seafood, and bakery items to the generalist. The generalist pays for the groceries, often at special Peapod counters in the back of the store. The order is then taken to a holding area in the supermarket or in a trailer, where the appropriate items are kept cold or frozen until the deliverer picks up a set of orders and takes them to the customers. At each stage— ordering, shopping, holding, and delivery—the processes are modularized to provide personalized service at a relatively low cost.

If a customer has a problem, he or she can call Membership Services, and a service representative will try to resolve the matter. Peapod treats each call as yet another opportunity to learn (and remember) each customer's prefer- ences and to figure out what the company can do to improve service for cus- tomers as a whole. For example, service representatives found that some customers were receiving five bags of grapefruits when they really wanted only five grapefruits. In response, Peapod now routinely asks customers to confirm orders in which quantities might be confused.

Peapod's results stand as a testament to the power of learning relation- ships. The four-year-old service, which has 7,500 customers and revenues of about $15 million, has a customer-retention rate of more than 80%. And the service accounts for an average of 15% of the sales volume of the 12 Jewel and Safeway stores where Peapod shops for its customers.

Vying for the End Customer

Retailers, insurance agents, distributors, interior decorators, building contractors, and others who deal face-to-face with the end customer can certainly make the case that they should be the ones who control the relationship with that customer. On the other hand, manufactur- ers and service providers have an advantage when a customer often buys the same type of product (toiletries, magazines, or office sup- plies); when products can be economically delivered to the home or office (personal computers, software, or services such as lawn care and plumbing); or when customers already value their relationship with the product or brand (as with premium Scotch, designer jeans, or lux- ury watches).

Obviously, those boundaries are permeable and constantly shifting: manufacturers and service providers can become retailers and vice versa. And advances in technology are making it increasingly easy for one member of the value chain to undermine another's natural ad- vantages. Consider, for example, three basic reasons consumers go to retail stores: to obtain the information they need to make a purchas- ing decision, to pay for the product, and to take possession of it.

Thanks to the same information-based technologies that make learning relationships possible, consumers increasingly will not have to visit stores for any of those reasons.

Today, consumers can get better information—information that is unbiased, comparative, accurate, and immediate—through on-line services, CD-ROM catalogs, and fax-response systems, and eventually they will be able to obtain it through interactive TV. As the continuing boom in catalog and home-TV shopping attests, consumers and organizations can buy goods and services over the phone and through dedicated on-line services as easily as, if not more easily than, in person, and security measures will almost certainly be in place soon that will make it possible to purchase products through the Internet. Finally, almost anything can be delivered direct to the home thanks to Federal Express, UPS, dedicated delivery services, and (for digitized products) fax and on-line services.

For retailers, the message is clear: if they want to maintain or increase their competitive advantage, they must begin establishing learning relationships with their best customers today. On the other hand, a manufacturer or a service company one or more links removed from the end user has a variety of options. It could build collaborative learning relationships with those occupying the next link, gaining knowledge about their wants, needs, and preferences over time, and mass-customizing products and services to meet their requirements. That is the approach ITT Hartford's Personal Lines business is taking with the independent agents who sell its automobile and home insurance. And it is also the direction in which Andersen—which realizes that individual home owners buy windows too infrequently to form a productive, long-term relationship with the company—is heading. Although Andersen plans to continue to mass-customize windows for consumers, it also intends to cultivate learning relationships with architects, home builders, and window distributors.

Another option for a manufacturer or a service company is to form tighter partnerships with retailers so that together they control the learning relationships with individual end customers. Such a partnership would require sharing information and knowledge (and maybe a common database), linking operations tightly so consumers' desires could be translated efficiently and quickly into tailored products and services, and possibly making joint investment and strategic decisions on how best to serve end customers over time. This option might make sense for companies such as automakers, which rely heavily on dealers to provide the touch, feel, and test drive necessary for consumers to make a buying decision.

How to Build Learning Relationships

If managers decide that their company can and should cultivate learning relationships with customers, how do they go about it? There are basically four components to think about: an *information strategy* for initiating dialogues with customers and remembering their preferences; a *production/delivery strategy* for fulfilling what the company learns about individual customers; an *organizational strategy* for managing both customers and capabilities; and an *assessment strategy* for evaluating performance.

THE INFORMATION STRATEGY

Cultivating learning relationships depends on a company's ability to elicit and manage information about customers. The first step is to identify those individual customers with whom it pays to have a learning relationship. That is easy for businesses like hotels or airlines, whose customers make reservations in their own names and whose transactions and preferences are easy to track.

In industries whose customers are anonymous, such as retailing, a company may have to use one of two approaches to persuade them to identify and provide information about themselves: show them that it can serve them better if they do or give them something of value in return, such as a gift or a discount. For example, Waldenbooks offers a 10% discount on all purchases if customers identify themselves by becoming Preferred Readers. The program allows the company to track the purchases of those customers at any Waldenbooks store. Learning about customer preferences enables the bookseller to let a particular customer know when, for example, the next William Styron novel will be out or when an author whose work the customer has purchased will be in a local store, signing books.

Few companies will want to have such relationships with all customers. Waldenbooks' program, for example, is aimed at people who spend more than $100 a year at its stores. As a screening device, the company charges a $10 annual fee for Preferred Reader status.

As with any new program, it is often best to begin with a company's most valuable customers. When the company sees that the value of a learning relationship with them exceeds the costs, it can gradually expand the program to other customers.

Once a company has identified the customers with whom it wishes to have a learning relationship, there are a number of ways in which it can conduct a productive dialogue. A rapidly expanding array of interactive technologies—including electronic kiosks, on-line services, and database-driven mail—are making such dialogues easier and less costly. (See "How to Interact: A Sampler of Today's Technologies.") Businesses that naturally involve personal contact with customers, either on the phone or in person, have golden opportunities to learn about them.

How to Interact: A Sampler of Today's Technologies

Interactive media that allow marketers to send specific messages to specific consumers and to conduct a dialogue with actual and potential customers already exist. One is the Internet, which now boasts more than 15 million users. Using it simply to prospect for customers remains problematic owing to the hostility of many users to commercial advertising on the Internet. But many companies have found the Internet to be a good way to obtain information from or about customers through bulletin boards, direct connections, and company-specific information services.

Other on-line services, such as those provided by Prodigy, America Online, and Compuserve, are much more advanced than the Internet in providing a full-fledged, structured medium through which customers and companies can interact. And several company-specific on-line services, such as grocery deliverer Peapod's, have proved useful for facilitating dialogues with customers.

Electronic kiosks have a wide variety of applications for interacting directly with customers. Some are purely informational—like those that provide directions to local spots from a hotel lobby. Others dispense coupons or gift certificates. And an increasing number are being used to dispense mass-customized products, including greeting cards, business cards, and sheet music.

A variety of interactive telephone services exists already. Seattle-based FreeFone Information Network offers one on the West Coast that enables marketers to find consumers willing to participate in a dialogue. When people sign up for the service, they fill out a questionnaire that is used to determine which advertiser's message is sent to which person. Each time a consumer makes a personal call and listens to a sponsored message while waiting for the call to connect, FreeFone credits the household account a nickel. The household gets a dime if the consumer requests more information, a coupon, or a telephone connection to the advertiser. Companies that advertise through FreeFone, including TicketMaster, the U.S. Postal Service, NBC, and the National Association of Female Executives, can learn a great deal about each house-hold. But FreeFone will not divulge a caller's identity to an advertiser unless the caller chooses to reveal it.

"Cash-back telephone coupons" provide a similar way for companies and consumers to learn about each other over the phone. These services, offered by such companies as Chicago-based Scherers Communications, are essentially reverse 900 numbers. For example, a car manufacturer might credit someone $5 for watching a videotape touting some particular models and calling in with the personal identification number contained on the tape.

Fax response is being used by many business-to-business organizations and a small but growing number of consumer-goods manufacturers to give customers up-to-the-minute price quotations and product options. Fax response provides the marketer with the telephone-number identity of the individual who requested the information, which can be linked with transactional data as well as with mailing information.

R.R. Donnelley & Sons' selective binding technology, which enables printers to put different pages in different editions of a given publication, has made it possible for publishers to mass-customize periodicals. *Farm Journal*, for example, assembles information on individual subscribers—how many acres of what particular crops they have planted, how many head of cattle they own, and so on—and then uses Donnelley's technology to tailor the editorial content and the advertising of each edition for the particular subscriber.

In conducting a dialogue with customers, it is important that the database "remember" not just preferences declared in past purchases but also the preferences that emerge from questions, complaints, suggestions, and actions.

The Ritz-Carlton hotel chain trains all its associates—from those on the front desk to those in maintenance and housekeeping—how to converse with customers and how to handle complaints immediately. In addition, it provides each associate with a "guest preference pad" for writing down every preference gleaned from conversations with and observations of customers. Every day, the company enters those preferences into a chainwide database that now contains profiles of nearly a half million patrons. Employees at any of the 28 Ritz-Carlton hotels worldwide can gain access to those profiles through the Covia travel-reservation system.

Say you stay at the Ritz-Carlton in Cancún, Mexico, call room service for dinner, and request an ice cube in your glass of white wine. Months later, when you stay at the Ritz-Carlton in Naples, Florida, and order a glass of white wine from room service, you will almost certainly be asked if you would like an ice cube in it. The same would be true if you asked for a window seat in a restaurant, a minibar with no liquor in your room, or a variety of other necessities or preferences that personalize your stay at the Ritz-Carlton.

By retaining such information, a company becomes better equipped to respond to suggestions, resolve complaints, and stay abreast of customers' changing needs. Many companies make the mistake of treating customers as if they were static entities rather than people whose preferences, lifestyles, and circumstances are constantly evolving and shifting.

Some managers may wonder whether customers will see requests for in-depth personal information as an invasion of privacy. Most people don't mind divulging their shopping habits, measurements, and friends' names and addresses if they believe they're getting something of value in return. Consumers' fears also will be assuaged if a company states unequivocally that it will jealously guard personal information, which any company building learning relationships must do. Unlike mass marketers, who buy and sell customer data willy-nilly, companies seeking to build learning relationships realize that such information is a precious asset.

THE PRODUCTION/DELIVERY STRATEGY

Children can create an unlimited number of unique designs with Lego building blocks. Service and manufacturing companies that have successfully mass-customized employ a similar approach: they create modules—components or processes—that can be assembled in a variety of ways to enable the companies to tailor products or services for specific customers at a relatively low cost. (See chapter 9 "Making Mass Customization Work.") Admittedly, there is more opportunity to adopt this approach in some businesses than in others. For example, the Ritz-Carlton is more of a customizer than a mass customizer. If it could figure out how to mass-customize its services, as Peapod has done, it would be able to cater to the preferences of more of its customers *and* increase its profits.

However, creating process or component modules is not enough. A company also needs a design tool that can take a customer's requirements and easily determine how to use its capabilities to fulfill them. Individual, Inc.'s SMART system and Andersen's Window of Knowledge system are examples of design tools that enable companies to be as effective as possible in ascertaining what customers need, as efficient as possible in production and delivery, and as effortless as possible in matching the two.

THE ORGANIZATIONAL STRATEGY

Traditional marketing organizations depend on product managers to push the product out the door, into the channels, and into customers' hands. Product managers are generally responsible for performing market research, specifying the requirements for a fairly standardized offering, and developing the marketing plan. And once the product is introduced, they are rewarded for selling as much of it as possible. While these techniques are ideally suited for mass marketing, they are ill suited for learning relationships in which the reverse is required: extracting a customer's wants and needs from a dialogue and creating the product or service that fulfills those requirements.

To build learning relationships, companies don't need product managers; they need *customer managers.* As the term implies, customer managers oversee the relationship with the customer. While they are responsible for a portfolio of customers with similar needs, they also are responsible for obtaining all the business possible from *each* customer, one at a time. To do this, customer managers must know their customers' preferences and be able to help them articulate their needs. They serve as gatekeepers within the company for all communication to and from each customer.

In addition, companies need *capability managers,* each of whom executes a distinct production or delivery process for fulfilling each customer's requirements. The head of each capability ensures that appropriate capacity exists and that the process can be executed reliably and efficiently.

Customer managers must know what capability managers can provide and must take the lead in determining when new capabilities may be required to meet customers' needs. For their part, capability managers must know what customer managers require and be able to figure out how to create it. For instance, when a Peapod customer informed his customer manager (a Membership Services representative) that he wanted to be able to order both ripe and unripe tomatoes, the company expanded the capabilities of its ordering software and created a new set of capability managers: produce specialists. These specialists have the skills and experience to squeeze tomatoes and thump melons, for example. Similarly, a customer manager at four-year-old Individual asked the company's manager of information suppliers—the capability manager responsible for managing and acquiring new sources of information—to add the *New England Journal of Medicine* after learning that a client needed articles from the publication. Individual expands the number of its sources by 75 to 100 per year in this manner.

In contrast to the traditional product manager's role of finding customers for the company's products, the role of the customer manager is finding products for the company's customers. Often, a customer manager will learn of a need for some product or service component that the organization does not consider itself competent to produce or deliver. The capability manager might then arrange to obtain it from a strategic partner or a third-party vendor. For example, it would not pay for AT&T's computer hardware and software business, AT&T Global Information Solutions (formerly NCR), to write software for every conceivable customer need. When a customer-focused team (the unit's equivalent of customer managers) learns that a customer needs a particular application that is unavailable in-house, it often asks a capability-management team to acquire or license the software.

In all cases, however, the customer manager must be held accountable for satisfying the customer. At ITT Hartford's Personal Lines business, every time a customer (an independent agent) makes a request, Personal Lines forms an instant team composed of people from whichever service modules (underwriting, claims payment, or servicing, for example) are needed to satisfy the request. But the customer manager is the one responsible for guaranteeing the promised customized service. He or she specifies the commitment to the agent at the beginning of each transaction, and a tracking system ensures that it is fulfilled.

THE ASSESSMENT STRATEGY

Obviously, the value of a learning relationship to the company will vary from customer to customer. Some customers will be more willing than others to invest the time and effort. Those willing to participate are going to have a wide variety of demands or expectations, meaning that the company will have a varying ability to contribute to and profit from each relationship. Companies should therefore decide which potential learning relationships they will pursue.

The ideal way to approach this task is to think about a customer's lifetime value. Lifetime value is the sum of the future stream of profits and other benefits attributable to all purchases and transactions with an individual customer, discounted back to its present value. In their article "Zero Defections: Quality Comes to Services" (HBR September–October 1990), Frederick F. Reichheld and W. Earl Sasser, Jr., showed that the longer customers are retained by a company, the more profitable they become because of increased purchases, reduced operating costs, referrals, price premiums, and reduced customer acquisition costs. We would add one more element to the list: some customers will have higher lifetime values because the insights they provide to the company may result in new capabilities that can be applied to other customers. Although it is a daunting task, companies seeking to build learning relationships should therefore try to track as many of those elements as they can, using such information as transactional histories and customer feedback.

A company's *customer share*—its share of each customer's total patronage—is one of the most useful measures of success in building a learning relationship. To calculate customer share, a company must have some idea of what the customer is buying from the competition and what he or she might be willing to buy from the company. The best source of such information is the customer—another reason why dialogue is critical. Yet another important performance measure is what we call *customer sacrifice:* the gap between what each customer truly wants and needs and what the company can supply. To understand individual customer sacrifice, companies building learning relationships must go beyond the aggregate customer-satisfaction figures that almost everyone collects today. That is why Peapod asks every customer at every shopping session how well it did on the last order. Understanding and tracking this gap will enable customer managers to demonstrate the need for new capabilities to deepen learning relationships and will give capability managers the information they need to decide how to expand or change their company's capabilities.

Becoming a Learning Broker

After a company becomes adept at cultivating learning relationships with its current customers, how might it expand? Two choices are obvious: acquire new customers in the company's current markets or expand into new locations. But there is a third option: deliver *other* products to *current* customers and become a learning broker.

Because Peapod's customers already know how to interact with its on-line ordering system, the company could easily broker new product categories. For example, if Peapod could gain entry into a chain of home-improvement centers (meaning on-line access to the chain's computerized list of stock-keeping units and prices, and Peapod shoppers' access to the stores themselves), its knowledge about its customers and its customers' knowledge about it would immediately transfer to a whole new set of "virtual aisles." And once again, it would be Peapod—not the chains or the manufacturers that supply them—that would control the relationship with the customers. By arbitraging the information between customers and companies that supply products and services that they could potentially use, Peapod would have become a bona fide learning broker.

Discussions of what life will be like in the information-rich, interactive future often focus on personal electronic "agents" that will watch out for each individual's information and entertainment needs, sifting and sorting through the plethora of channels, messages, and offerings. But the dynamics of learning relationships are such that learning brokers can provide that service today in a wide variety of domains. They could provide individual customers with products and services beyond those that their companies have traditionally supplied. They also could advise their customers about other offerings and be on the lookout for items they might want.

One of the best examples of a company that already serves its customers in this fashion is the United Services Automobile Association. Seventy years ago, USAA began providing automobile insurance to military officers. It now supplies its customers—whom it still limits to current and former military officers and their families—with a wide variety of products and services. They include all types of insurance, full-service banking, investment brokerage, homes in retirement communities, and travel services. USAA also offers a buying service through which it purchases and delivers other companies' products, including automobiles, jewelry, major appliances, and consumer electronics. The relationship with the customer, however, remains the sole dominion of USAA.

USAA members have learned over the years that the company stands behind everything it sells and looks after their best interests. As more than one member has said, USAA could sell almost anything to them. More than nine of every ten active-duty and former military officers are members. And since opening up its services to members' adult children in the 1970s, USAA has been able to attract more than half of them, showing that learning relationships can even span generations.

The role of a learning broker clearly makes sense for distributors or agents such as Peapod and Individual, two companies that make no products themselves. Such companies are relatively free to go to whatever company can provide exactly what their customers want and need. Whether to take the path of a learning broker is a more complex decision for a manufacturer or a service company. But it is not out of the question. A company can become a hybrid like USAA: it offers its members a wide variety of other companies' products, but, in its core business, financial services, it offers only its own products. While it may be difficult to imagine today, many companies could eventually decide that it pays to become a learning broker even of competitors' products. But adopting that strategy will make sense only if a company reaches the point where its knowledge of its customers and their trust in it yield a greater competitive advantage and greater profits than merely selling its own products can. When that happens, learning relationships with end customers will have become the company's primary competency.

5

Is Your Company Ready for One-to-One Marketing?

Don Peppers, Martha Rogers, and Bob Dorf

Practiced correctly, one-to-one marketing can increase the value of your customer base. The idea is simple: one-to-one marketing (also called relationship marketing or customer-relationship management) means being willing and able to change your behavior toward an individual customer based on what the customer tells you and what else you know about that customer. Unfortunately, too many companies have jumped on the one-to-one bandwagon without proper preparation. The mechanics of implementation are complex. It's one thing to train a sales staff to be warm and attentive; it's quite another to identify, track, and interact with an individual customer and then reconfigure your product or service to meet that customer's needs.

So is your company ready to implement a one-to-one marketing program? In large part, the answer depends on the scope of the program. For some companies, being ready simply means being prepared to launch a limited initiative. Substantial benefits can be gained from taking steps—even small ones—toward one-to-one marketing in specific functional areas. For others, being ready means being positioned to implement an enterprise-wide program. To help you assess the type of program you should begin with—and determine what you need to do to prepare—we've developed a list of activities and a series of exercises designed for executives, managers, and employees at all levels in your company, as well as for your customers and channel

This article is adapted from the authors' book The One to One Fieldbook: The Complete Toolkit for Implementing a 1 to 1 Marketing Program *(Currency/Doubleday, 1999).*

partners. Reviewing the list and working through the exercises will help you determine what type of one-to-one marketing program your company can implement immediately, what you need to do to position it for a large-scale initiative, and how to prioritize your plans and activities.

Why One-to-One?

Before determining the correct scope of your company's one-to-one marketing efforts, you need to understand the rationale for undertaking a one-to-one initiative and the basic components of such a strategy. Relationship marketing is grounded in the idea of establishing a learning relationship with each customer, starting with your most valuable ones. (See chapter 4, "Do You Want to Keep Your Customers Forever?") Think of a learning relationship as one that gets smarter with each interaction. The customer tells you of some need, and you customize your product or service to meet it. Every interaction and modification improves your ability to fit your product to this particular customer. Eventually, even if a competitor offers the same type of customization and interaction, your customer won't be able to enjoy the same level of convenience without taking the time to teach the competitor the lessons your company has already learned.

There are four key steps for putting a one-to-one marketing program to work: identifying your customers, differentiating among them, interacting with them, and customizing your product or service to fit each individual customer's needs.

IDENTIFYING YOUR CUSTOMERS

To launch a one-to-one initiative, your company must be able to locate and contact a fair number of its customers directly, or at least a substantial portion of its most valuable customers. It's critical to know customers in as much detail as possible: not just their names and addressable characteristics (such as addresses, phone numbers, or account codes), but their habits, preferences, and so forth. And not just a snapshot—a onetime questionnaire. You need to recognize the customer at every contact point, in every medium used, at every location, and within every division of your company, no matter which product line is involved. Remember, however, that the "customers" who benefit from your one-to-one program may not be limited to the end users of your product or service. If, for example, you are a manufacturer selling to retailers, then you will also want to apply the principles of one-to-one marketing to create better relationships with your channel members and other intermediaries in your demand chain.

DIFFERENTIATING YOUR CUSTOMERS

Broadly speaking, customers are different in two principal ways: they represent different levels of value and they have different needs. Once you identify your customers, differentiating them will help you to focus your efforts so as to gain the most advantage with the most valuable customers. You will then be able to tailor your company's behavior to each customer in order to reflect that customer's value and needs. The degree and type of differentiation in a company's customer base will also help you decide on the appropriate strategy for a given business situation.

INTERACTING WITH YOUR CUSTOMERS

Improving both the cost-efficiency and the effectiveness of your interactions with customers is a critical component of a one-to-one marketing program. Cost-efficiency improves by directing customer interactions toward more automated and therefore less costly channels. For example, a company that provides helpful, up-to-date information at its Web site won't need to spend as much as it once did supporting a more expensive call center. Effectiveness improves by generating timely, relevant information, providing either better insight into a customer's needs or a more accurate picture of a customer's value. Every interaction with a customer should take place in the context of all previous interactions with that customer. A conversation should pick up where the last one left off, whether the previous interaction occurred last night or last month, at the call center or on the company Web site.

CUSTOMIZING YOUR ENTERPRISE'S BEHAVIOR

Ultimately, to lock a customer into a learning relationship, a company must adapt some aspect of its behavior to meet that customer's individually expressed needs. This might mean mass-customizing a manufactured product, or it could involve tailoring some aspect of the services surrounding a product—perhaps the way the invoice is rendered or how the product is packaged. In any case, the production or service-delivery end of your business has to be able to treat a particular customer differently based on what was learned about that customer by the sales, marketing, or any other department. In rushing to reap the rewards of relationship marketing, it's easy for a business to overlook this critical fourth step, leading many to misunderstand the entire discipline as simply an excuse for direct mail and telemarketing. (See Susan Fournier, Susan Dobscha, and David Glen Mick, "Preventing the Premature Death of Relationship Marketing," HBR January–February 1998.)

These four implementation steps overlap considerably. Nevertheless, they are roughly in order of increasing complexity and increasing benefit for a company. Identifying and differentiating customers, the first two steps, are largely internal "analysis" steps, whereas interacting with your customers and customizing products and services are external "action" steps, visible to the consumer. From that perspective, the four steps can be used as a kind of general checklist to guide your efforts in implementing a one-to-one marketing program. If you can't identify your individual customers, you have no hope of differentiating them, much less adapting your behavior to address each one's needs.

Starting Small

Many managers dismiss the possibility of one-to-one marketing because they feel it is an unattainable goal. And yes, it's true that there are numerous reasons to think twice before launching a full-scale program. For one, your company's information technology department may be too swamped—or not sufficiently developed—to handle all the tasks that one-to-one marketing demands. Maintaining a customer database, having one system communicate seamlessly with another, tracking each customer's contacts with the company—all of those activities require IT development, direction, and support. In addition, one-to-one marketing requires a certain amount of capital investment across the board, and many companies are unwilling or unable to provide enough funding to all relevant areas to make the initiative worth the effort. And, of course, there are organizational puzzles to be solved. It's easy to assign responsibility for a product, but who takes responsibility for developing a customer relationship across different business units? Which business unit "owns" the customer, anyway?

Those are serious considerations—and they don't even scratch the surface of the scope of cultural change a one-to-one marketing strategy may demand. But even a very modest one-to-one initiative—one that affects just one area, such as your sales force or your call center or your Web site—can produce substantial benefits. Besides hinting at the value of a full-scale program, often these short-term results are themselves enough to justify the funding required for an incremental effort. Among them:

- **Increased cross-selling.** A retail bank, for instance, that is able to in- crease the average number of accounts per customer from 1.8 to 2.5 will enjoy a very significant, and measurable, financial benefit. If you can track just a few of your business's transactions, you can compare the amount of added benefit you're getting from cross-selling and up- selling. You ought to see higher unit margins as well, provided you're tracking this metric on a per-customer basis.

- **Reduced customer attrition.** One of the primary, and early, benefits of a one-to-one marketing program is that it generates increased loy- alty among customers. Try tracking defections among customers ex- posed to a relationship-marketing initiative compared with a statisti- cally identical control group not exposed to the initiative. What would it be worth to your business, just in terms of reduced acquisition costs, for instance, to increase average customer tenure by 10%? Or what about increasing the average likelihood of repurchase by 10%?

- **Higher levels of customer satisfaction.** Granted, this is a "soft" rat- ing. But it's easily measured and can provide quick support of one of the benefits of a relationship-marketing program. To get at *real* cus- tomer satisfaction, you might want to measure your customers' "likeli- hood to recommend" your product, or something more tangible than most traditional customer-satisfaction indices.

- **Reduced transaction costs and faster cycle times.** One-to-one marketing is basically oriented around making it increasingly conve- nient for a customer to buy, which translates directly into a more efficient organization. The fewer things a customer has to specify each time business is done, the more efficient the transaction will be.

Exhibit 5-1   Getting Started

The following activities are keyed to the four steps of a one-to-one marketing program: identifying customers, differentiating among them, interacting with them, and customizing your product or service to meet each customer's needs. Most companies should be able to accomplish these activities fairly readily. If you have not yet identified your end-user customers, you can apply these suggestions to your channel partners. At some point, however, you will need to identify and interact with your end-user customers to get the most out of your relationship-marketing program.

IDENTIFY

Activity	Steps to consider
Collect and enter more customer names into the existing database	• Use an outside service for scanning or data entry. • Swap names with a noncompetitive company in your field.
Collect additional information about your customers	• Use drip-irrigation dialogue: ask your customers one or two questions every time you are in touch with them.
Verify and update customer data and delete outdated information	• Put your customer files through a "spring cleaning." • Run your database through the National Change of Address (NCOA) file.

DIFFERENTIATE

Activity	Steps to consider
Identify your organization's top customers	• Using last year's sales or other simple, readily available data, take your best guess at identifying the top 5% of your customers.
Determine which customers cost your organization money	• Look for simple rules to isolate the bottom 20% of your customers (such as customers who haven't ordered in more than a year or those who always bid you out) and reduce the amount of mail you send them.
Select several companies you really want to do business with next year	• Add them to your database, and record at least three contact names per company.

Find higher-value customers who have complained about your product or service more than once in the last year	• Baby-sit their orders: put a product or quality-assurance person in touch with them ASAP to check on your progress.
Look for last year's large customers who have ordered half as much or less this year	• Go visit them now, before your competition does.
Find customers who buy only one or two products from your company but a lot from other businesses	• Make them an offer they can't refuse to try several more items from you.
Rank customers into A, B, and C categories, roughly based on their value to your company (Don't try to isolate the top 5% or bottom 20%—any "blunt instrument" criterion such as annual spending or years doing business with the company will work.)	• Decrease marketing activities and spending for the C's and use the savings to fund increased activities for the A's.

INTERACT

Activity	Steps to consider
If you are focusing on channel members, call the top three people at your top 5% of customers	• Don't try to sell—just talk and make sure they are happy.
Call your own company and ask questions; see how hard it is to get through and get answers	• Test eight to ten different scenarios as a "mystery shopper." Record the calls and critique them.
Call your competitors to compare their customer service with yours	• Repeat the above activity.
Use incoming calls as selling opportunities	• Offer specials, closeouts, and trial offers.
Evaluate the voice response unit at your customer information center	• Make the recordings sound friendlier, be more helpful, and move customers through the system faster.
Follow the interaction paper trail through your organization	• Seek to eliminate steps: reduce cycle times to speed up your response times to customers.
Initiate more dialogue with valuable customers	• Print personalized messages on invoices, statements, and envelopes. • Have sales reps sign personal letters rather than mass-mailing letters signed by a senior manager.

	• Have the right people in your organization call the right customer executives. (That is, have your CIO call another CIO, or have the VP of marketing call the business owner.) • Call every valuable customer your company has lost in the last two years and give them a reason to return.
Use technology to make doing business with your company easier	• Gather the e-mail addresses of your customers in order to follow up with them. • Offer alternative means of communication. • Consider using fax back and fax broadcast systems. • Scan customer information into database.
Improve complaint handling	• Plot how many complaints you receive each day and work to improve the ratio of complaints handled on the first call.

CUSTOMIZE

Activity	Steps to consider
Customize paperwork to save your customers time and your company money	• Use regional and subject-specific versions of catalogs.
Personalize your direct mail	• Use customer information to individualize offers. • Keep the mailings simple.
Fill out forms for your customers	• Use laser equipment to save time and make you look smarter.
Ask customers how, and how often, they want to hear from you	• Use fax, e-mail, postal mail, or personal visits as the customer specifies.
Find out what your customers want	• Invite customers to focus groups or discussion meetings to solicit their reactions to your products, policies, and procedures.
Ask your top ten customers what you can do differently to improve your product or service	• Respond to their suggestions. • Follow up and repeat the process.
Involve top management in customer relations	• Give them lists of questions to ask based on the history of individual customers.

In Exhibit 5-1 "Getting Started," we've listed activities that most companies can accomplish fairly readily in each of the four key implementation areas of relationship marketing. Some of these activities may work for your business today; others may stimulate your thoughts for future initiatives. If at least some of the ideas make sense to you, then you should generate your own list and begin to identify the most important activities to put into practice. In any case, the activities we've listed here make some of the more basic concepts of customer-relationship management practical, and they can help your company take the first step.

Assessing Your Situation

If you are considering launching a full-scale one-to-one initiative, the first critical task is to assess your current situation. How big is the gap between where you are and where you want to be? What should you work on first? Where are you in relation to your competitors? The exercises entitled "The Broad View" and "The One-to-One Gap Tool" will give you a clearer picture of just how much work lies before you.

The principal groups whose evaluations would be helpful are:

- anyone involved in conceiving or developing one-to-one marketing strategies at your company;
- all senior managers;
- a selection of relevant middle managers and field managers;
- channel members, if applicable;
- line employees who interact directly with customers (sales and service people, call service representatives, and retail clerks);
- a representative group of your customers.

Responses will probably vary widely among those groups, and that is exactly the way the exercises should be evaluated—by comparing how different groups rate your company. It is important to value the answers in inverse proportion to the respondents' position in the corporate pecking order. The best lessons, in fact, will likely come from hearing what your customers have to say about your company's readiness to engage in one-to-one relationships with them. Beyond that, however, there is enormous value in comparing the answers given by different constituencies, in different locations and functions, facing different agendas and issues. Being relatively concise, "The Broad View" can be widely circulated to ensure breadth of coverage. "The One-to-One Gap Tool" will capture a more robust analysis of how your company sees itself—both culturally and organizationally.

The Broad View

Is Your Company Ready for One-to-One Marketing?

Administer this exercise to a number of people at different levels and in different areas of your company. Also ask a representative group of customers to participate, although you'll need to tailor the language appropriately. There is enormous value in hearing what your customers have to say about your company's readiness to engage in one-to-one relationships with them as well as in comparing the answers from the various constituencies.

Note: If it seems impossible to identify even your most valuable end-user customers, then your best option may be to concentrate on developing relationships with the intermediaries in your demand chain whose identities you can readily acquire. If so, then complete this exercise (and "The One-to-One Gap Tool") as if it applied to channel members. But keep in mind that sooner or later you will almost certainly want to deal with end users themselves, even if your relationships with them will always include channel members.

1. **How well can your company identify its end-user customers?**
   a. We don't really know who our end-user customers are or how much business they give us.
   b. We have some information about our end-user customers in various files and databases around the company, but we're not sure what proportion of customers this really represents.
   c. Some of our business units know many of their customers' individual identities, but not all the units do. Our business units don't have a central database of customer identities, and we don't share much data on customers with one another.
   d. We sell to businesses or organizations, and although we know the identities of all or most of these organizations, we don't really know the individual players at each business.
   e. We know who most of our customers are individually, but we don't know much about their relations with one another. If a customer refers another customer to us, we don't track this in our database. And if a customer moves from one location to another, our database might show this as a customer defection followed by a customer acquisition.

f. We know most of our customers individually, and we can track them from place to place, division to division, or store to store.

2. **Can your company differentiate its customers based on their value to you and their needs from you?**

   a. Because we don't have much, if any, information on our customers' identities, naturally we are unable to differentiate our customers either by their value or by their needs.

   b. We have no real knowledge of how to rank our customers individually by their long-term value to us.

   c. We have an idea about how to calculate our customers' individual long-term value to us, but we don't have enough data to generate a reliable ranking of individual customers based on this calculation.

   d. We have identified a number of different needs-based segments of our "most valuable customers" (MVCs), but we don't have a reliable way of mapping particular customers into the right segment.

   e. We know how to rank most of our customers individually by value, and we can also identify, at least for most of our MVCs, their appropriate needs-based segment.

3. **How well do you interact with your customers?**

   a. We have no practical mechanism for interacting with our customers on an individual basis.

   b. We interact with some of our MVCs through personal sales calls and other forms of contact, but we don't systematically record these interactions through sales force automation or contact management systems. We rely instead on the initiative and memories of our account directors, salespeople, and others to manage these customer interactions.

   c. We interact with most of our MVCs through sales calls and other forms of contact, and we maintain fairly good records of those interactions and contacts in an automated system or customer database.

   d. We have direct interaction through mail, phone, or on-line media with a small proportion of our customers, but we don't coordinate these interactions across media.

   e. We interact through mail, phone, or on-line communication with all or a substantial number of our customers, and we coordinate the dialogue we have with any single customer across these different media.

4. **How well does your company customize its products and services based on what it knows about its customers?**

   a. We provide standard products and services and tailor few, if any, aspects of our behavior to the needs of individual customers.

b. We offer a range of options for our customers so they can choose specific product features for themselves, but we don't track or remember which features each customer chooses.

c. In the case of our MVCs, we sometimes customize our peripheral services—contract terms, billing formats, delivery modes, pallets and packaging, service options, and so forth—and we track each MVC's preferences.

d. We have modularized at least a few aspects of our core product or our peripheral services or both, and by configuring these modules in different ways, we can produce a variety of product-service combinations fairly cost-efficiently. For a substantial number of our customers, we track and remember which customers choose which options, so when a customer repeats a purchase with us we can automatically configure our product to that customer's previously stated preferences.

e. We have modularized many aspects of our core product or our peripheral services or both, and we can render a wide variety of product-service configurations cost-efficiently. Rather than asking customers to sort through all the options themselves, we interact with most or all of them to help them specify their needs. Then we map each customer into a particular needs-based category, propose a particular product configuration for that customer, and remember it the next time we deal with that customer.

The One-to-One Gap Tool

How Far Do You Have to Go to Prepare?
This exercise, to be administered to employees at various levels and in various functions, is designed to capture a robust analysis of how your company sees itself both culturally and organizationally. It should also be given to a representative group of customers, with the language tailored appropriately, in order to expose the gap between internal and external perceptions.

For each question listed below, select the statement that most closely reflects your opinion of the company as you view it today—not as you think it should be or it might be in the future.

Processes
I. **Does the company have established quality-assurance processes?**
   a. We do not consider quality-management practices.
   b. We would like to have formal quality-management initiatives.
   c. We have some methods in place to ensure quality-management initiatives.
   d. We have a formal quality-management organization.

2. Are the company's business processes customer-centric?

    a. We do not pay any attention to how our customers and our business processes interact.

    b. We have some understanding of the link between customers and our business processes.

    c. We understand most of the interactions between customers and our business processes.

    d. We have a full understanding of all the possible interactions between customers and our business processes.

Technology

3. Does the company take customers' needs into consideration when selecting and implementing technology?

    a. Our IT department is rather autonomous and in charge of technology acquisition.

    b. We ensure that our customers' needs, not just our own internal needs, are considered when selecting technology.

    c. We use some degree of customer validation when selecting technology.

    d. We ensure that all technology selections are customer-centric. For instance, we research how to improve customer convenience.

4. Does the company provide its employees with technology that enables them to help customers?

    a. We are not particularly advanced when it comes to technology.

    b. We encourage the use of technology that helps our daily interactions with customers.

    c. We provide technology in many areas to improve our daily interactions with customers.

    d. We provide the most effective technology available to all employees who interact with customers.

Knowledge Strategy

5. Does the company maintain a strategy for collecting and using information about customers?

    a. We handle information about customers poorly.

    b. We encourage the collection and use of information to gain knowledge about customers.

    c. We have programs to collect information and use our knowledge about select customers.

    d. We continually enhance our strategy to collect and use our knowledge about our customers.

6. **How effectively does the company combine information on customers with its experiences to generate knowledge about its customers?**

    a. We have poorly developed and inadequate processes for combining our data on customers with our own experience and views.

    b. We encourage using processes and systems that support the collection of both customer information and experiences about some customers.

    c. We have implemented systems and processes that collect and combine information and experiences about selected customers.

    d. We have rigorous processes to combine information and experiences about each customer.

Partnerships

7. **How does the company select its partners?**

    a. We pay little or no attention to whether the partners we select are customer-centric.

    b. We try to select partners that are customer-centric.

    c. We evaluate strategic partners based on their customer centricity.

    d. We evaluate all potential partners based on their customer centricity.

8. **Does the company understand the relationships among its customers and partners?**

    a. We have little or no understanding of the relationships among our customers and our partners.

    b. We try to understand the relationships among our customers and our partners.

    c. We understand the relationships among our customers and our partners.

    d. We understand and use the relationships among our customers and our partners.

Customer Relationships

9. **How effectively does the company differentiate its customers?**

    a. We do not differentiate among customers.

    b. We try to differentiate among customers.

    c. We collect and use information gleaned from interactions with customers to differentiate each customer and to evaluate the importance of each relationship.

d. We have a continuously updated customer-knowledge database that provides all the critical business information about our relationships with individual customers.

10. **What steps has the company taken to improve the total experience of its customers?**
    a. We pay little or no attention to the total experience of customers.
    b. We know all the points where customers are in contact with the business, and we manage these areas effectively.
    c. We conduct frequent surveys with selected customers and make improvements based on their feedback.
    d. We have a continual dialogue with each customer and use well-developed methods to improve our relationships.

11. **How effectively does the company measure and react to customers' expectations?**
    a. We make no effort to understand our customers' expectations.
    b. We have some idea of our customers' expectations and use them in building relationships.
    c. We periodically solicit customers' input about expectations and take actions to improve the relationships where possible.
    d. We work as a team with our customers to ensure that their expectations are met or exceeded.

12. **How effectively does the company understand and anticipate customers' behavior?**
    a. We pay little or no attention to the behavior of our customers.
    b. We understand the trends and buying patterns of our customers and consider them when making critical decisions.
    c. We collect data on our customers' preferences and other behaviors and use that information in our business planning.
    d. We maintain a profile of each customer and refer to it when dealing with customers.

Employee Management

13. **To what degree are employees empowered to make decisions in favor of the customer?**
    a. We encourage employees to strictly follow procedures and policies developed by top managers.
    b. We encourage employees to make independent decisions within the guidelines set by management.

c. We strongly encourage employees to make decisions that positively affect our customers' satisfaction.

d. We require every employee to take whatever action is appropriate to ensure the ultimate satisfaction of the customer.

14. **Has the company formally linked employees' rewards with customer-centric behavior?**

a. We make no link between employees' rewards and their treatment of customers.

b. We use ad hoc methods to reward customer-centric behavior.

c. We make customer-centric behavior a part of performance appraisal criteria.

d. We make customer-centric behavior a significant part of performance appraisal criteria.

Competitive Strategy

15. **To what extent does the company understand how customers affect the organization?**

a. We attach little significance to the views and opinions of customers.

b. We place some importance on understanding the impact of our customers on the business.

c. We place importance on understanding how a select group of customers affects our business.

d. We place vital importance on understanding how each customer affects our business.

16. **How much influence do customers' needs have on the company's products and services?**

a. We pay little or no attention to the needs of our customers when we design our products and services.

b. We attempt to develop products and services that meet our customers' needs.

c. We use input from selected groups to assist with the development of products and services.

d. We design products and services to meet the needs of individual customers.

17. **How effectively does the company build individualized marketing programs?**

a. We build all marketing programs to reach a mass market.

b. We build all marketing programs to fit a perceived niche market.

    c. We build some marketing programs that are specific to each customer's needs.

    d. We build all marketing programs to be specific to each customer's needs.

18. **How aware is the company of other organizations' approaches to building customer relationships?**

    a. We pay no attention to the customer-centric strategies of other companies.

    b. We know which companies are customer-centric regardless of the industry.

    c. We know how our competition approaches customer centricity.

    d. We know the best-in-class approaches to customer centricity.

Setting Priorities

Once you evaluate your business's ability to begin implementing a customer-relationship-management program, the next step is to think through the one-to-one marketing issues that are most (or least) relevant to your own particular business situation in order to prioritize your efforts. For instance, what role should a Web site play in your company's plan to create better relationships with customers? Instead of designing a Web site, should you try to customize your product offering, perhaps by lining up strategic alliances with businesses that provide complementary services?

To set priorities, you need to consider how differentiated your customer base is in terms of customer needs and value. You also need to determine how capable your company is at interacting and customizing—the two "action steps" in the four-step implementation process. For instance, if the needs of your customers vary widely, then you probably should focus first on customization: the more your customers' needs vary, the more attractive they will find a learning relationship.

Consider the contrast between a bookstore and a gas station. If a customer who enters a bookstore is reminded that his favorite author has a new book out that has been reserved in case he wants it, he is likely to become very loyal to the store. People's tastes in books vary widely, so it's a real service for a proprietor to remember a particular customer's preferences.

But if the same person pulls into a gas station and the attendant announces he has a new shipment of 93-octane in, just the way he knows the customer likes it, the "service" of remembering customer preferences is not nearly so attractive.

When customers' value to the enterprise vary widely, the top customers account for the vast majority of the business. We call this a *steep skew*. Relationship marketing is more cost-efficient for businesses with a steep skew than with a shallow one. The greater the skew, the more feasible it is to cultivate relationships with the most valuable customers. If the top 2% of your customers generate 50% of your profit, you can protect 50% of your bottom line by fostering learning relationships with just the top few customers. But if the top 20% of your customers make up 50% of your profit, then it will be ten times as costly to achieve the same bottom-line benefit.

There are specific strategies—we call them *migration strategies*—appropriate for dealing with a customer base that is not well differentiated in terms of needs or value. These strategies involve either expanding the definition of customer needs or value, or improving the interaction or customization capabilities of the enterprise.[1] For example, when dealing with a customer base that is characterized by a shallow value skew, one strategy would be to improve the cost-efficiency of the company's capability to interact—adding a call center, for instance, or a Web site. Basically, the less costly it is to interact, the less important it is to reserve interactions for top customers only.

Consider the example of the bookstore again. Although it might have a customer base with widely different needs, it also has a fairly flat value skew. Very few bookstore proprietors actually remember the preferences of their individual customers because it is simply not cost-efficient to do so. Even if the proprietor could remember the tastes of her top 100 customers—and teach her salespeople to do so as well—these top customers would probably account for less than 10% of the proprietor's business.

But if the bookstore were to increase its capability to interact cost-efficiently—say, by adding a Web site—then the amount of effort required to remember customers' preferences would decrease dramatically. Amazon.com really does remember its customers' individual tastes—and not just for a top few, but for thousands. As a result, Amazon.com is able to create the kind of learning relationship with its customers that will keep them loyal. It will always be at least a little easier for the customer to return to Amazon.com to find a book than to go someplace else and explain his preferences anew. This is one reason 59% of Amazon.com's sales come from repeat customers—roughly twice the rate of typical bricks-and-mortar bookstores.

Of course, Amazon.com and other companies that do such a good job of remembering their customers' tastes need to allow them to adjust their preferences freely. For instance, a customer might buy a book as a gift for someone with different interests, or he might simply change his mind about topics he's interested in.

A Foundation for the Long Term

It's been our experience that implementing even a relatively limited one-to-one program tends to motivate a company to take a more integrated, enterprise-wide view of its customers. String a few of these projects together and pretty soon you'll have to deal with a series of enterprise issues. People in different functions and from different business units will be working together more frequently on an ad hoc basis. Managers on one project will be trying to relate their metrics to the outcomes of other projects.

As you begin to take a more integrated view of the enterprise, certain organizational issues will arise. Considering the following questions may give you some advance warning:

- If you measure a customer's value across more than one division, will one person be in charge of that customer's relationship? If so, how will this be structured?
- Should the enterprise set up or modify its key-account selling system?
- Should the enterprise underwrite a more comprehensive information system, standardizing customer data across every division?
- Should the company be thinking about investing in a data mart or a data warehouse?
- Should the sales force be better automated? If so, who should set the strategy for how sales representatives interact with individual customers?
- Does it still make strategic sense to have different sales forces for different divisions?
- Is it possible for the company's various Web sites and call centers to work together better? That is, do Web pages coming from different

divisions and locations make sense to a customer looking at them in total? Can callers be "hot connected" between call centers? Should you combine your call centers?

- Should the company package more services with the products it sells? If so, how should those services be delivered?
- Should the business seriously explore investing in mass-customization manufacturing technologies?

To be ready to tackle issues like these, you might want to create a few programs now. Set up a multi-department committee to agree on a standard way to report customer information, for instance. Agree on a cross-divisional standard format for customer service calls. Come up with a weighted measure to rank customers by their overall value—not just their worth to one division.

Obviously, it's impossible to simply "strap on" a one-to-one marketing campaign and continue to do business in a traditional manner. Many companies—Dell, USAA, American Express, and Amazon.com, for instance—will be more successful at creating learning relationships with their customers because their businesses were built on the basis of direct customer interaction.

But it's also possible to make steady, incremental progress by concentrating on the four implementation steps and applying them to different aspects of your current business. Large, well-established enterprises like Pitney Bowes, Wells Fargo, 3M, Owens Corning, British Airways, and Hewlett-Packard have begun creating stronger, more interactive relationships with their customers. They implement these strategies piece by piece, in one business unit at a time, wrestling with one obstacle at a time. But they are making progress and gaining a significant competitive advantage as a result.

So instead of asking, "Is your company ready to implement a one-to-one marketing strategy?" perhaps the better question is, "How much of a one-to-one marketing program is your company ready to deal with today?" Clearly, putting relationship marketing to work properly involves much more than simply sending out personalized mail, training your call center personnel in phone etiquette, or designing a user-friendly Web site. But when correctly executed, the process of making even incremental progress toward becoming a one-to-one marketer can pay immediate dividends as you strengthen and deepen your company's relationships with its customers.

Note

1. For a more comprehensive explanation of these concepts, see Don Peppers and Martha Rogers, *Enterprise One to One: Tools for Competing in the Interactive Age* (New York: Currency/Doubleday, 1997), especially chapter 3 and chapter 13.

PART

III

The Business of
Mass Customization

6
Breaking Compromises, Breakaway Growth

George Stalk, Jr., David K. Pecaut, and
Benjamin Burnett

When is a mature, slow-growth business not a mature business? How do rapidly growing companies emerge from stagnant, dead-in-the-water industries? The station-wagon segment of the North American auto market was dying when, in 1984, Chrysler Corporation introduced the minivan. Over the next ten years, minivan sales grew eight times faster than did the industry overall. For the last 15 years, the do-it-yourself home-improvement business as a whole has grown barely 5% per year while Home Depot has racked up 20% growth. Over capacity and flat demand plague the airline industry, but that hasn't kept Southwest Airlines Company from growing seven times faster than the industry average over the past decade.

What senior managers at Chrysler, Home Depot, and Southwest have in common is the wisdom, curiosity, and perseverance to explore the compromises their industries were forcing customers to endure. And each acted on the insight that breaking those compromises would release enormous trapped value—enough to stimulate major sales and profit growth. (See Exhibit 6-1 "Breakaway Growth: Compromise Breakers Have Outperformed Their Industries.") In fact, the concept of breaking compromises is one of the most powerful organizing principles we have seen for companies that wish to achieve breakaway growth.

Compromises are not trade-offs. Trade-offs are the legitimate choices customers make among different product or service offerings. Trade-offs typically come from fundamental differences in cost structures that are usually reflected in prices. With products, the trade-offs

Exhibit 6-1 Breakaway Growth: Compromise Breakers Have Outperformed Their Industries

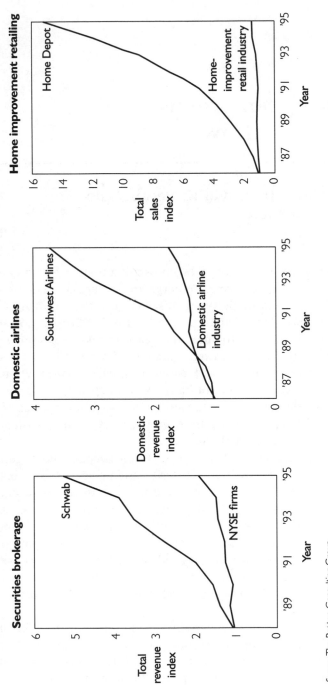

Source: The Boston Consulting Group

often arise from differences in design or in the cost of materials. With textiles, for example, there is a trade-off between price and quality because better fabrics tend to have higher thread counts. In service, trade-offs are common because delivering greater convenience or customization often entails higher cost. Thus taxi service costs more than bus service, and a meal delivered by room service costs more than the same meal ordered in the hotel restaurant.

A compromise, in contrast, is a concession demanded of consumers by all or most service or product providers. Whereas tradeoffs let customers choose their preference among alternatives, compromises offer no choice. Trade-offs allow different offerings to appeal to different segments; compromises benefit no particular segment. Trade-offs are very visible; most compromises are hidden.

In picking a hotel room, for instance, a customer can *trade off* luxury for economy by choosing between a Ritz-Carlton and a Best Western. But the hotel industry forces customers to *compromise* by not permitting check-in before 4 P.M. Similarly, until recently, most auto dealers forced a compromise on customers by not offering weekend repair and maintenance services. There is no law of nature that says that cars can't get fixed on weekends or that hotel rooms can't be ready before late afternoon. Compromises occur when an industry imposes its own operating practices or constraints on customers, leaving them no choice. It's the industry's way or no way. And often customers accept compromises as the way the business works.

Henry Ford's famous car in any color—as long as it's black—is one type of compromise. Such a compromise denies customers the selection they want. Or customers are forced to wait. Today's car buyer can custom-order virtually any car if the selection on the dealer's lot is inadequate, but the industry will make customers wait six to eight weeks for delivery. In other situations, customers may be forced to use a high-cost service or to pay a premium to get the quality they want. Because the family washing machine can't safely handle all fabrics, customers have to spend extra time and money on dry cleaning. The compromise often becomes visible when customers have to modify their behavior to use a company's product or service. Until recently, dishwashers did a satisfactory job of washing the dishes, but they made enough noise to wake the dead. Their owners had to arrange a time when they were out of hearing range to wash the dishes.

Compromises creep into businesses in various ways. Some, like hotel check-in times, are imposed by standard operating practices that no one questions. Others stem from conscious decisions that may make marginal economic sense—as long as customers adjust their behavior. For example, it may make sense for a supplier to deliver only once a week, but doing so forces customers to hold inventory between deliveries. The most important compromises, however, are forced on customers simply because companies have lost touch with those customers' needs. Finding and breaking those compromises can unleash new demand and create breakaway growth.

The Great Pasta Compromise

Contadina, an operating unit of Nestlé, has created a high-growth business by breaking the compromises imposed on consumers of pasta. Contadina's fresh pasta product is sold in supermarkets, cooks in minutes in boiling water, and comes in many varieties, including ravioli and tortellini.

Before Contadina's innovation, consumers faced one trade-off and a multitude of compromises in their quest to eat pasta. The trade-off was between eating pasta at home—where someone has to make it—and eating out at a restaurant. Pasta at home is less expensive. The restaurant has more variety and means less work, but it costs more.

The great pasta compromise begins after the decision to stay home and make it yourself. Homemade pasta is inexpensive and fresh. But making pasta from scratch is time-consuming and difficult. The first product to try to break the compromise was dry pasta. Dry pasta costs more than homemade pasta and it is not as fresh, but it is much easier and faster to make.

The next run at the great pasta compromise was frozen pasta. Frozen pasta, which often comes in a microwavable container, is even quicker and easier to cook than dry pasta and requires little cleanup. But frozen pasta costs more than either homemade or dry pasta, and it is often less tasty.

In the mid-1980s, Contadina made its run at the great pasta compromise with the introduction of a fresh pasta product. Contadina's fresh pasta is twice the price of dry pasta and comes in a smaller package that doesn't serve as many people. It is five times more expensive per serving than dry pasta. Why, then, do people buy Contadina? Consumer research provides some interesting insights. Naturally, consumers like its freshness and its ease of cooking. More surprising is the fact that consumers are choosing Contadina over a meal at a restaurant. Before Contadina's fresh pasta became available, many people said they would never eat tortellini or ravioli at home, because preparing them from scratch was just too much trouble.

Breaking the great pasta compromise not only has made it easier to prepare good-tasting pasta at home, it also has upset the old trade-off between eating at home and going to a restaurant. In this context, Contadina makes a lot of sense. Consumers get the taste, variety, and freshness that can be found at restaurants, but in a product they can cook and eat at home for less money.

Often, when segmentwide compromises imposed on consumers are broken, traditional trade-offs are sidestepped and fundamental changes in the definition of the business occur. This usually means a dramatic shift in the set of relevant competitors. Because compromise breakers often find themselves competing against companies that are higher cost and higher priced, they are often able to grow rapidly and profitably by gaining share from their new set of rivals. Contadina grew at high double-digit rates to become a leader in fresh pastas and sauces by the 1990s, with hundreds of millions of dollars in sales.

A Breakthrough for Car Buyers

Breaking compromises between an entire industry and its customers can release tremendous value. Circuit City, best known in the United States as a big-box consumer-electronics and appliance retailer, is a successful company, with sales growing at 26% per year and earnings at 30%. Its one major problem is that it is about to run out of real estate. After opening stores in virtually all major markets in the United States, Circuit City needs to go somewhere else for fast-growth opportunities.

The retailer has found what it believes to be a promising opportunity in an unlikely place—the used-car business. In October 1993, Circuit City launched CarMax, a company whose strategy is to revolutionize the way used cars are sold in the United States.

Selling used cars is a business with a stigma. In the past, most people who bought used cars couldn't afford new ones. The automakers, who naturally wanted to sell new cars, reinforced the stigma. When Chrysler introduced its successful K-car in the early 1980s, Roger Smith, then chairman of the board of General Motors Corporation, was asked how GM would respond to the threat. Smith belittled the K-car by saying that "General Motors' answer to the Chrysler K-car is a two-year-old Oldsmobile."

This attitude toward used cars has not changed much. In the summer of 1995, a *Business Week* journalist grilled the program manager for the new Ford Taurus about the car's price. In frustration, the program manager responded that the 1996 Taurus was priced to sell 400,000 units a year. "If Joe Blow can't afford to buy a new car . . . let him buy a used car" (*Business Week,* July 24, 1995).

The used-car business may get no respect, but it should. Annual used-car sales in North America top $200 billion, making used cars the third-largest consumer category after food and clothing. In fact, there are more sales of used cars and light trucks than of new ones, and demand for used cars is growing faster. Moreover, the quality of used cars has risen with the rise in the quality of new cars.

Despite improvements in product quality, the business of selling used cars is virtually unchanged. A customer who opts for a used car faces many compromises. First, the buyer has to locate a car, usually by reviewing the classified advertisements in the local paper. Product variety is limited. In Toronto, for example, 20 to 30 used Tauruses are advertised for sale in the newspaper at any one time—from individuals, from dealers specializing in used cars, and from new-car dealers who also sell used cars. In the case of private sales, the buyer must call, make an appointment, and hope the seller will actually be there at the appointed time. The buyer must drive to see the car—which is unlikely to turn out to be the one the buyer wants or in good condition or priced reasonably or even still there: it could already have been sold.

When the buyer finds an attractive car, he or she can't expect to see any maintenance records. Some dealers certify their cars, but in Ontario, for example, certification means only that the glass is not cracked, that the lights and brakes work, that the exhaust does not leak, and that the tires have sufficient tread. In other words, certification guarantees only the bare necessities for roadworthiness.

Buyers of used cars, then, risk ending up owning a car with mechanical problems. Beyond this, they must endure a time-consuming and truly horrific buying process—more accurately, up to four processes: finding and buying the car, financing it, insuring it, and selling the old car. Buyers are at a disadvantage because knowledge about the product is asymmetrical: the seller knows more than the buyer. Often the buyer is subjected to high-pressure sales tactics and forced to haggle over the price with salespeople whom he or she suspects are dishonest. And should problems arise, there is no clear recourse for the buyer.

The managers of Circuit City observed the size and growth of used-car sales and saw that many of the distinguishing capabilities of their own consumer-electronics business could break the compromises imposed on buyers of used cars.

Circuit City is known for its high variety of merchandise. CarMax takes the same approach. A typical large used-car dealer has only about 30 vehicles in stock. A large new-car dealer who sells used cars might have 130 vehicles. The first CarMax, in Richmond, Virginia, had 500 cars. The two stores that opened in Atlanta in August 1995 have 1,500 each.

CarMax further enhances customer choice by harnessing Circuit City's considerable systems capabilities. At CarMax, customers have access to computerized information through a kiosk that enables them to sort through the inventory of cars available not only at that site but at all the stores in the region. When CarMax advertises in any of the Richmond or Atlanta papers, it advertises inventory from both locations.

Unlike Circuit City, CarMax does not keep its inventory indoors. There is only one vehicle on display in the showroom, and it is fitted with arrows pointing to the 110 spots that have undergone performance and safety checks. The showroom's computerized kiosks provide information on the vehicles in stock, including their location on the lot. Should a customer be shopping with the family and want to see and drive a particular vehicle, CarMax provides a supervised day-care center for the children.

CarMax uses professional uniformed sales representatives, whose first job is to explain how to use the kiosk and then to help customers find the car they want. CarMax prefers *not* to hire people with experience in selling new or used cars. Instead, it wants to hire presentable people whom it can train for two weeks (compare that with the minimal or nonexistent training that employees receive at new- and used-car dealers) and pay a set dollar amount per vehicle regardless of its selling price. This is an interesting departure from Circuit City's practice of paying a percentage-of-sales commission that encourages aggressive "selling up." CarMax did not want that pressure on its customers, so it designed an incentive system that eliminates the pressure on its sales representatives.

CarMax sets prices at below the average Blue Book value and offers no-haggle pricing and no-hassle guarantees. Every CarMax vehicle comes with the 110-point safety check and a 30-day warranty. For some cars, warranties of up to four years are available. In addition, CarMax customers have a five-day return guarantee: the car may be brought back with no questions asked as long as it has not been driven more than 250 miles.

Financing is available from NationsBank Corporation or from Circuit City's financing arm. Circuit City's financing tends to be for a longer term and usually requires lower deposits. Progressive Insurance will insure both the vehicle and the driver on the spot. People buying cars from CarMax can sell their old cars as well. The sale of the used car is a separate transaction from the purchase of a car. CarMax will buy any used car—although not at a price everyone will accept.

The jury is still out on the success of CarMax. A host of imitators have emerged. Circuit City does not divulge the performance of its CarMax unit, but 4 stores were opened in 1995 and 90 more are planned by the year 2000. The race is on. (See Exhibit 6-2 "Driving Bottom-Line Growth at CarMax.") Both used- and new-car dealers are likely to be bloodied. Historically, new-car dealers have sold about 80% of used cars that are less than four years old—CarMax's core offering. And those sales have accounted for anywhere from 35% to 65% of the dealers' profitability overall.

In addition, the sales of new cars are at risk. A popular saying in the automotive industry is that when you buy a new car and drive it off the lot, what you own is a very expensive used car. On average, the value of a new car plummets 28% in the first week after its sale. At CarMax, it is not uncommon to find current model-year vehicles with low mileage at substantially lower prices than those of new vehicles. In breaking so many of the compromises imposed on used-car buyers, CarMax may end the old trade-off between buying a used car and buying a new one.

CarMax is not the first to try to break the compromises imposed on used-car buyers. In northern New Jersey, there is a used-car dealership that has tackled the variety compromise by putting 600 cars on its lot and giving customers more choice. But this dealership has left

Exhibit 6-2 Driving Bottom-Line Growth at CarMax

CarMax's innovative formula makes it possible to turn inventory much faster than is possible for traditional dealerships, thereby driving growth of the bottom line as well as growth of revenue.

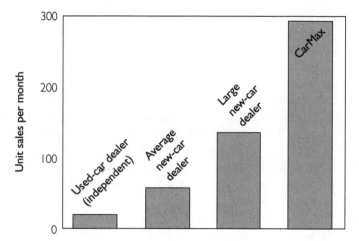

everything else the same. Customers still have to haggle, obtain their financing and insurance from somewhere else, and dispose of their old car. Other dealers are touting no-haggle pricing. What sets CarMax apart is that it has put it all together: CarMax sells variety, it sells value, it sells convenience, it promises that you can be in and out in 90 minutes with a car, and it delivers a comfortable experience. Many of the car dealers near CarMax locations are matching CarMax on price, and they think, mistakenly, that the job is done. It's not. People want a different buying experience and they're getting it from CarMax.

Finding Opportunities in Any Business

Growth strategies built around the idea of breaking compromises are neither new nor limited to a few particular industries. But to visualize such a strategy requires a company's managers to clear their heads of the conventional thinking that pervades their industry. The Charles

Schwab Corporation has grown steadily over two decades by breaking one industry compromise after another.

Schwab began as a discount stockbrokerage in 1975, when U.S. equity markets were deregulated and price competition on stock-trading commissions was introduced. Discount brokers ended a major compromise for individual investors, who had hitherto been forced to put up with high prices if they wanted to buy and sell securities.

Schwab, however, saw that the discount brokerage segment was itself imposing new compromises. Customers who opted for a low price worried about service reliability. Schwab tackled the problem head-on, first by investing heavily in computer technology that allowed almost immediate confirmation of orders over the telephone. At the time, even Merrill Lynch & Company could not do that. Schwab also invested in the firm's brand name and in retail offices, both of which instilled confidence in consumers. In the process, the firm broke the compromise between price and reliable service and grew dramatically through the early 1980s.

Schwab saw that other compromises remained to be broken. In exchange for low prices, customers had been compromised on convenience, flexibility, and ease of transferring funds. In the early 1980s, Schwab pioneered 24-hour-a-day, seven-day-a-week service. It introduced the Schwab One cash-management account with Visa card and checking privileges, copying a Merrill Lynch product but eliminating the need to deal with a full-commission broker. Schwab also pioneered automated phone trading and eventually electronic trading directly from the customer's personal computer.

Over time, Schwab's management realized that the company was no longer a simple discount broker but in fact a broad, value-priced provider of cash, stocks, bonds, and mutual funds. The compromises Schwab had broken had generated a 20% to 25% per year growth rate and made Schwab the largest non-full-commission broker in the United States. But Schwab was ready to break yet another compromise to fuel its next stage of growth.

Until 1992, most consumers wanting to buy mutual funds had been forced to choose among different fund companies, each of which serviced its own accounts. Because diversification and high performance were not easily accomplished within a single family of funds, many consumers placed money with a number of different fund-management companies. Most investors were frustrated by the complexity of dealing with different statements, different rules, and different sales representatives.

In 1992, Schwab changed the scenario by introducing OneSource, a single point of purchase for more than 350 no-load mutual funds in 50 different fund families. OneSource gives customers a single account with one monthly statement that tracks the performance of all their funds. There are no transaction fees on OneSource accounts, so customers can shift their money among different fund families without any charge. Schwab can do this because it is paid directly by the funds as their sales representative and subaccount processor.

OneSource has grown to include 500 mutual funds, driving Schwab's mutual-fund assets from $6 billion in 1991 to more than $60 billion in 1996 and making it the third-largest mutual-fund distributor in the United States. No longer forced to compromise on assortment, price, and convenience, consumers have been flocking to OneSource to manage their investments.

Schwab's experience illustrates that relentless breaking of compromises can be a source for continuing growth. In fact, there are at least seven ways in which companies can find and exploit compromise-breaking opportunities in any industry.

Shop the way the customer shops. At Schwab, the most important source of ongoing insight is employees who use the company's products and services just as Schwab's customers do. For example, the belief that customers would value the convenience of 24-hour-a-day, seven-day-a-week systems was heavily supported by Schwab's own employees, who wanted that kind of flexibility in managing their own investments. Unfortunately, in many industries, executives never know how customers shop. In the auto industry, executives of the Big Three do not buy cars. Their secretaries do it for them, over the telephone. The cars are delivered to the executives clean, full of gas, and ready to go. For most Big Three executives, buying a car the way ordinary customers do would be an out-of-body experience.

Pay careful attention to how the customer really uses the product or service. In all industries, people exhibit *compensatory behaviors*. They devise their own ways of using the product or service to compensate for the fact that if they did only what the company intended them to do, they wouldn't really get what they wanted. In every product category, consumers can undertake dozens of compensatory behaviors, and each of those can have significant compromise-breaking potential.

In the brokerage business, it was common knowledge that customers often called back a second or even a third time to confirm that their trade had gone through at the price they had requested. Schwab paid careful attention to customers' actual behavior and realized that the ability to provide immediate confirmation at the time an order was taken would eliminate those second and third calls—saving customers a lot of trouble and giving Schwab a significant advantage over other brokers.

Explore customers' latent dissatisfactions. Most companies ask their customers to describe their dissatisfactions with existing products and services. Such surveys usually lead to helpful improvements, but truly significant breakthroughs are generally the result of tapping into much deeper dissatisfactions. Those can be called *latent dissatisfactions* because consumers are unable to articulate their unhappiness with the product or service category. Chrysler's development of the minivan, for example, tapped into latent dissatisfaction with both station wagons and full-size vans. Station wagons couldn't carry enough and were hard to load and unload. Full-size vans were more useful, but they were not fun to drive. Minivans broke the compromise by "cubing out" the box design of the station wagon. Ford Motor Company and GM had both researched customers' feelings about station wagons and had found that they could meet obvious needs with features such as two-way doors, electric rear windows, and third seats. But they did not explore the "white space" between station wagons (based on car platforms) and vans (based on truck platforms). The minivan—a van based on a car platform—was hidden in this white space defined by customers' latent dissatisfactions.

Look for uncommon denominators. Over time, companies tend to drift toward providing products or services that, on average, meet the needs of large numbers of customers. But compromises often lurk in this common-denominator approach. Schwab, for example, has now separated the service channel for the high-volume equity trader from that for the ordinary investor, whose needs are simpler. Each receives different services and pays different fees.

Some companies are reluctant to abandon the approach of averaging costs across all customers, because they believe abandoning it will reduce the profitability of their high-volume accounts. But recent history suggests that if managers don't separate out what should be discrete businesses, a new or existing rival will do it for them. Recognition of that fact begets a relentless search for new compromises to break.

Pay careful attention to anomalies. Anomalies often are a rich source for compromise breaking. The one regional sales office that significantly outperforms all others and for which there is no obvious explanation; the factory that appears to have a scale disadvantage but still has a lower production cost; the supplier who has lower cost and higher quality despite having an older product design: those anomalies are all worth exploring as potential compromise-breaking opportunities.

In Schwab's case, the idea of creating local offices grew out of an anomaly. Charles Schwab's uncle was looking for a business to run and Schwab decided to open an office in Sacramento, California, to give his uncle something to do. At the time, offices were seen as unnecessary and costly overhead for discount brokerage firms.

Subsequently, Schwab noticed that Sacramento was significantly outperforming other cities that had no offices. There was no obvious explanation. By exploring this anomaly carefully, Schwab discovered that retail sales offices had a number of important advantages even for a firm that typically dealt with customers by telephone. Local offices provided a rich source of customer leads through walk-in traffic and reassured those new customers who had concerns about trusting a broker they had heard of only from television. The offices provided a sense of solidity and a place customers could go to transact business. Schwab discovered that even in a high-tech age, customers like knowing that there is an office down the street or at least across town. Fully probing this anomaly led Schwab to build a large retail network.

Look for diseconomies in the industry's value chain. Today, in industry after industry, companies are innovating the management of their value chain in ways that are more rewarding for consumers. When the Schwab firm entered the mutual fund business, its first thought was to create its own family of funds. Careful analysis of the industry value chain, however, revealed a bigger opportunity. Only a handful of the largest companies had sufficient economies of scale to distribute their funds cost-effectively—and those companies lost the ability to talk directly to their individual customers. Schwab's solution was to become an intermediary between its own customer base and a large number of subscale mutual-fund companies. Through OneSource, the firm served the needs of the fund companies and at the same time interposed itself between the funds and the customer. Schwab's ownership of the direct customer relationship can now provide a platform for growth in other financial services, such as insurance.

Look for analogous solutions to the industry's compromises. Some of the best compromise-breaking ideas are probably already out there—in someone else's industry. Circuit City's CarMax borrows many practices from other retail sectors. For example, its idea of offering extended warranties on used cars is borrowed from appliance and consumer-electronics retailing. To keep inventory moving and selection fresh, CarMax has copied a practice used commonly in soft-goods retailing—automatically discounting inventory as it ages. CarMax's practice of offering flat sales commissions and its low-pressure selling tactics can be observed in a number of sectors. Best Buy Company, one of Circuit City's competitors in electronics retailing, uses that approach to create the kind of low-key, self-serve environment CarMax was looking for.

An Organizing Principle for Growth

Many companies today are searching for growth. But how and where should they look? Managers will often turn first to line extensions, geographic expansion, or acquisitions. In the right circumstances, each of those makes sense. But we believe that innovations that break fundamental compromises in a business are far more powerful.

Breaking compromises can, in fact, provide an organizing principle for the pursuit of growth. The CEO of a large financial-services company asked his initially skeptical management team to specify and value all the compromises imposed on its customers. The exercise was eye opening.

To get its employees to focus their energies on compromise breaking, a company should start by asking them to immerse themselves in the customer's experience. It is critical to develop a strong, almost visceral feel for the compromises consumers experience. Whirlpool Corporation, the $8 billion appliance maker, identified a specific individual who personified the compromises all its customers bore.

Whirlpool's market research showed consumers to be generally satisfied with the home appliances they owned. But digging deeper, Whirlpool discovered a reservoir of latent dissatisfaction with all the activities for which the appliances were used—doing laundry, preparing food, cleaning up after meals. Although consumers didn't expect a lot more of their washing machines, ranges, and dishwashers, they were nevertheless very dissatisfied with household chores.

Those latent dissatisfactions became the basis for Whirlpool's brand strategy. In 1992, after decades of competing mostly on cost with companies such as General Electric, Whirlpool wanted to build a new and more profitable strategy around a more sharply differentiated brand. Management knew it needed to articulate the strategy and mobilize all employees behind a vision. Someone at Whirlpool saw an interview on a national television-news program with an overworked woman named Gail and taped it, recognizing Gail as the embodiment of Whirlpool's target customer. Gail was a 40-year-old woman taking care of several children at home while holding down a full-time job. Gail did all the cooking, the laundry, and the housework. Her husband's role was apparently restricted to playing sports with the children and helping them with their homework. The image was consistent with Whirlpool's research, which showed that women in the United States who work as many hours as their husbands in jobs outside the home continue to do most of the household chores as well. Gail personified the pressed-for-time working woman.

At the end of the video clip, the interviewer turns to her and says, "You're taking care of everyone in this family. Who takes care of you?" Before she can reply, her husband answers for her, "I take care of Gail." Gail shoots him a look that could kill.

The video, which became a rallying point for Whirlpool's new strategy, challenged all employees to think about how Whirlpool could be the company that takes care of Gail. Why, for example, was it taking Gail so long to clean up after meals? The traditional stove top was obviously designed by someone who was spared the daily responsibility of keeping it clean. The top of Whirlpool's CleanTop stove is completely flat, eliminating all the grease traps of the old design. Dishwashers used to be deafening, but now Gail can work on her kitchen computer while Whirlpool's Quiet Partner dishwasher is running.

More compromises wait to be broken. Why is doing the laundry such a chore? Gail's washing machine takes less time than her dryer to complete its cycle. Gail compensates by starting with lighter, faster-drying loads first. But eventually the process bogs down, and Gail is wasting time and energy running to the basement because no one makes a synchronized washer and dryer.

Breaking compromises can be a powerful organizing principle to enlist an entire organization in thinking about growth. The lesson from all the high-growth compromise breakers we've observed is this: The opportunity to identify and exploit compromises for faster growth and improved profitability is there for the taking. But managers must go to customers and look for themselves. This isn't a job that can be delegated to the market research department. Managers must ask themselves why customers behave the way they do. An auto dealer told us how proud he was of his expansive, well-lit lot with no fences. "My customers like to come after hours to look at cars and trucks," he proclaimed. He apparently never asked himself why they would do such a strange thing. And it never occurred to him that they might be looking at cars after hours precisely because they didn't want to have to deal with *him*.

To find the kinds of growth opportunities companies like CarMax, Schwab, and Contadina are pursuing, managers have to get inside the customer's skin and ask, What compromises am I putting up with? What's wrong with this picture? Where's the minivan in this company?

7

The Four Faces of Mass Customization

James H. Gilmore and B. Joseph Pine II

Virtually all executives today recognize the need to provide outstanding service to customers. Focusing on the customer, however, is both an imperative and a potential curse. In their desire to become customer driven, many companies have resorted to inventing new programs and procedures to meet every customer's request. But as customers and their needs grow increasingly diverse, such an approach has become a surefire way to add unnecessary cost and complexity to operations.

Companies throughout the world have embraced mass customization in an attempt to avoid those pitfalls and provide unique value to their customers in an efficient manner. Readily available information technology and flexible work processes permit them to customize goods or services for individual customers in high volumes and at a relatively low cost. But many managers at these companies have discovered that mass customization, too, can produce unnecessary cost and complexity. They are realizing that they did not examine thoroughly enough what kind of customization their customers would value before they plunged ahead with this new strategy. That is understandable. Until now, no framework has existed to help managers determine the type of customization they should pursue.

We have identified four distinct approaches to customization, which we call *collaborative, adaptive, cosmetic,* and *transparent.* When designing or redesigning a product, process, or business unit, managers should examine each of the approaches for possible insights into how best to serve their customers. In some cases, a single approach will dominate the design. More often, however, managers will discover that they need a mix of some or all of the four approaches to serve their own particular set of customers.

Defining the Four Approaches

Let's summarize what characterizes the approaches and the conditions under which each should be employed.

Collaborative customizers conduct a dialogue with individual customers to help them articulate their needs, to identify the precise offering that fulfills those needs, and to make customized products for them. The approach most often associated with the term *mass customization*, collaborative customization is appropriate for businesses whose customers cannot easily articulate what they want and grow frustrated when forced to select from a plethora of options.

Paris Miki, a Japanese eyewear retailer that has the largest number of eyewear stores in the world, is the quintessential collaborative customizer. The company spent five years developing the Mikissimes Design System (to be called the Eye Tailor in the United States), which eliminates the customer's need to review myriad choices when selecting a pair of rimless glasses. The system first takes a digital picture of each consumer's face, analyzes its attributes as well as a set of statements submitted by the customer about the kind of look he or she desires, recommends a distinctive lens size and shape, and displays the lenses on the digital image of the consumer's face. The consumer and optician next collaborate to adjust the shape and size of the lenses until both are pleased with the look. In similar fashion, consumers select from a number of options for the nose bridge, hinges, and arms in order to complete the design. Then they receive a photo-quality picture of themselves with the proposed eyeglasses. Finally, a technician grinds the lenses and assembles the eyeglasses in the store in as little as an hour.

Adaptive customizers offer one standard, but customizable, product that is designed so that users can alter it themselves. The adaptive approach is appropriate for businesses whose customers want the product to perform in different ways on different occasions, and available technology makes it possible for them to customize the product easily on their own.

Consider the lighting systems made by Lutron Electronics Company of Coopersburg, Pennsylvania. Lutron's customers can use its systems to maximize productivity at the office or to create appropriate moods at home without having to experiment with multiple switches each time they desire a new effect. Lutron's Grafik Eye System, for example, connects different lights in a room and allows the user to program different effects for, say, lively parties, romantic moments, or quiet evenings of reading. Rather than repeatedly having to adjust separate light switches until the right combination is found, the customer can quickly achieve the desired effect merely by punching in the programmed settings.

Cosmetic customizers present a standard product differently to different customers. The cosmetic approach is appropriate when customers use a product the same way and differ only in how they want it presented. Rather than being customized or customizable, the standard offering is *packaged* specially for each customer. For example, the product is displayed differently, its attributes and benefits are advertised in different ways, the customer's name is placed on each item, or promotional programs are designed and communicated differently. Although personalizing a product in this way is, frankly, cosmetic, it is still of real value to many customers. (Witness the billions of dollars that consumers spend each year on such products as embellished T-shirts and sweatshirts.)

The Planters Company, a unit of Nabisco, chose cosmetic customization when it retooled its old plant in Suffolk, Virginia, in order to satisfy the increasingly diverse merchandising demands of its retail customers. Wal-Mart wanted to sell peanuts and mixed nuts in larger quantities than Safeway or 7-Eleven did, and Jewel wanted different promotional packages than Dominick's did. In the past, Planters could produce only long batches of small, medium, and large cans; as a result, customers had to choose from a few standard packages to find the one that most closely met their requirements. Today the company can quickly switch between different sizes, labels, and shipping containers, responding to each retailer's desires on an order-by-order basis.

Transparent customizers provide individual customers with unique goods or services without letting them know explicitly that those products and services have been customized for them. The transparent approach to customization is appropriate when customers' specific needs are predictable or can easily be deduced, and especially when customers do not want to state their needs repeatedly. Transparent customizers observe customers' behavior without direct interaction and then inconspicuously customize their offerings within a standard package.

Consider ChemStation of Dayton, Ohio, which mass-customizes a product that most of its competitors treat as a commodity: industrial soap for such commercial uses as car washes and cleaning factory floors. After independently analyzing each customer's needs, ChemStation custom-formulates the right mixture of soap, which goes into a standard ChemStation tank on the customer's premises. Through constant monitoring of its 80-to-1,000-gallon tanks, the company learns each customer's usage pattern and presciently delivers more soap before the customer has to ask. This practice eliminates the need for customers to spend time creating or reviewing orders. They do not know which soap formulation they have, how much is in inventory, or when the soap was delivered. They only know—and care—that the soap works and is always there when they need it.

Challenging the Mass-Market Mind-set

Although each of the four companies has implemented a strikingly different customization strategy, all share an orientation that challenges the conventional concept of markets and products. As mass production took hold in the hearts and minds of managers during the past century, the definition of a market shifted from a gathering of people for the sale and purchase of goods at a fixed time and place to an unknown aggregation of potential customers. Today as markets disaggregate, the definition is changing again: customers can no longer be thought of as members of a homogeneous market grouping. In fact, the concept of markets needs to be redefined still further as customization becomes more commonplace. (See "Gaining Access to New Markets.")

Gaining Access to New Markets

As the concept of a mass market gained currency a century ago with the success of such giants as Sears, A&P, Coca-Cola, and Ford, all too many managers lost sight of a simple fact known for ages by every butcher, cobbler, and corner grocer: every customer is unique. Economies of scale in manufacturing and distribution brought down the price of mass-produced goods so much that all but the most well-to-do customers were often willing to forgo their individuality and settle for standardized—but very affordable—goods.

Still, the uniqueness of individual customers never went away; it was just subsumed in the averages of countless bell curves in every market-research study ever performed. The concept began coming back into view when companies discovered segmentation in the 1950s and niche marketing in the 1980s. The rise of mass customization in the 1990s has been both a response to and, with the pioneers' success, the impetus behind the now commonplace notion of *segments of one:* every customer is his or her own market segment with specific requirements that must be fulfilled. And so it seems that we have come to the end of a 100-year progression.

Or have we? In fact, the journey does not end with every customer being his or her own market. The next step, a widespread recognition that multiple markets reside within individual customers, will turn the entire notion of markets and customers completely inside out.

The idea that every customer is in different markets at different times and different places is not as heretical as it initially might sound. For instance, newspaper publishers have long recognized that most of their customers have more leisure time on Sundays to read the paper and accordingly have filled that edition with a greater number and wider variety of stories. Similarly, airlines, hotels, and car-rental companies find that the desires of their clients differ greatly depending on whether they are traveling for business or for leisure—and differ yet again when they combine the two. One executive at a major airline remarked, "We've even found that the needs of business travelers differ depending on whether they are going to or coming from a meeting." In the apparel industry, a given customer could be in the market for casual wear at one time and for business attire at another. And with "casual Fridays" becoming increasingly common, many people must at least on occasion enter that new market known as "business casual."

Indeed, acknowledging that individual customers constitute multiple markets gives new meaning to the term *market* that approximates its original conception: the bringing together of a customer and a provider to fulfill that customer's unique needs as they exist at the present time and under the current circumstances. Only those companies that take their approach to customization down to this level will gain access to the multiple markets within each of us.

How can companies tackle this task? If the technological wherewithal exists, the easiest approach would be to design a product that could adapt to whatever market its user happened to be in—such as a car transmission that can be sporty for tooling down the coast or smooth for taking the in-laws out to dinner. For frequently purchased goods and services, a company could work with individual customers first to identify the markets they potentially could be in at different times and in different circumstances and then to maintain a distinct profile for each possibility. News providers, for example, could collaborate with customers to understand how much news and what type each wanted to read depending not only on the day of the week but also on that day's particular events and on each customer's constantly rotating field of interest. An airline could likewise maintain subprofiles that highlight each customer's changing preferences (for instance, preferred drink when going to a meeting—Pepsi with lime; preferred drink when coming from a meeting—Scotch on the rocks).

A real opportunity arises here because even the customers themselves may not realize these distinctions. Many people in many situations will discover things about themselves only in a collaborative dialogue with a trusted supplier. Together, customer and supplier will create the multiple markets within.

Instead of focusing on homogeneous markets and average offerings, mass customizers have identified the dimensions along which their customers differ in their needs. These points of *common uniqueness* reveal where every customer is not the same. And it is at these points that traditional offerings, designed for average requirements, create *customer sacrifice gaps:* the difference between a company's offering and what each customer truly desires.

To be effective, mass customizers must let the nature of these sacrifice gaps drive their individual approaches to customization. Paris Miki understood that consumers rarely have the expertise to determine which eyeglass design best fits their facial structure, desired look, and coloring, and therefore chose to collaborate with customers to help identify their largely unarticulated needs. Lutron adopted adaptive customization because it knew that no two rooms have the same lighting characteristics and that both individuals and groups use any given room in multiple ways. Planters realized that each of its retail customers varied in how it wanted to receive and merchandise standard peanuts, so cosmetic customization was its favored choice. And ChemStation understood that although each of its customers had unique formulation and delivery needs, none of them wanted to be bothered with either the day-to-day procedures or the formulation of such a mundane part of its business as soap.

Altering the product itself for individual customers provides the most clear-cut means of customization. But adept mass customizers realize that customizing the actual product is only one way to create customer-unique value. Customizing the *representation* of the product—or how it is presented or portrayed to the customer—can be effective as well. In fact, separating the product from its representation can provide a useful framework for considering which forms of customization are most appropriate for a given business. (See Exhibit 7-1 "The Four Approaches to Customization.")

A cosmetic customizer changes only the representation of the product—the packaging in the case of Planters. Collaborative customizers change the product itself in addition to changing some aspect of the representation. Paris Miki changes both the eyewear and its digitized representation—the shape and placement of the eyeglasses on the customer's on-screen image, the display of information about the particular lens, and the adjectives used to describe the desired look. A transparent customizer uses a standard representation to mask the customization of the product. ChemStation's standard storage and dispensing tank, emblazoned with the company's logo, conceals the fact that ChemStation customizes the soap and its delivery. Finally, adaptive customizers change neither the product nor the representation of the product for individual customers; instead, they provide the customer with the ability to change both the product's functionality and its representation to meet his or her particular needs. Each Lutron customer programs a lighting effect by adjusting bars that represent the intensity of each light in the room; the customer then can label the particular lighting effect.

Companies customize representations when they use design tools such as the Mikissimes Design System to alter their products' descriptions. The following components also can change the form of an offering for individual customers:

Packaging: containers for shipment; bar codes, labels, and other materials-handling information; instructions; and storage and dispenser features.

Marketing materials: sales brochures, flyers, videotapes, and audiotapes; and client references and customer testimonials.

Exhibit 7-1   The Four Approaches to Customization

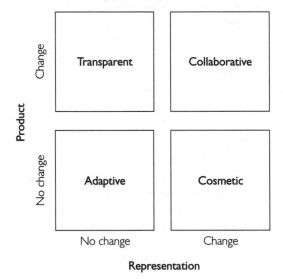

Placement: where, when, how, and to whom the product is delivered; position while on display or in use; and delivery frequency and special handling procedures.

Terms and conditions: purchase price; payment and discount terms; promotions, warranties, and guarantees; ordering policies; and after-sale service procedures.

Product names: brand names; cobranding (the presentation of two brands together); club memberships; and privileges for frequent customers, such as frequent flier programs.

Stated use: advertised purpose and operability; and perceived advantages, conveniences, or other benefits to the user.

Choosing the Right Approach

The four companies that we focus on identified the critical customer sacrifice gaps in their businesses and then carefully identified not only what but also when to customize in an effort to create the greatest customer-unique value at the lowest possible cost. Paris Miki customizes during the design of eyeglasses, whereas Planters customizes on packing lines during production; ChemStation customizes during both production and delivery, whereas people customize Lutron's product during each individual use. Instead of taking a hit-or-miss approach, these four companies customized only where it counted.

Let's explore how to determine which types of customization are appropriate for a given business.

COLLABORATIVE CUSTOMIZATION

The customer's inability to resolve trade-offs on his or her own has led Paris Miki and other companies in industries as diverse as apparel, windows, news services, and industrial valves to collaborative customization. Customers in these industries have to make one-time decisions based on difficult and multidimensional trade-offs—trade-offs such as length for width, comfort for fit, or complexity for functionality. This either/or sacrifice gap built into the onetime decision points toward the need to work directly with individual customers in order to determine together the customized goods or services they require. Customizing the representation permits customers to participate in the design stage and play with the possibilities available to them.

Take the customer's struggle to find the right eyeglasses. Paris Miki decided that the best way to help customers discover their unknown needs and resolve the inherent trade-offs associated with buying glasses was to allow each one to explore and manipulate a digitized representation of the potential final product. With this sophisticated design tool, trained opticians now assist customers in discovering the perfect, unique look that they would not otherwise have identified or found.

Collaborative customization also works effectively in the shoe industry. Many buyers of mass-produced shoes have to sacrifice a perfect fit on one foot to avoid a fit that is too tight or too loose on the other. Furthermore, no matter how broad the selection is in a traditional shoe store, customers have to make trade-offs among a number of superficial design elements when selecting a pair of shoes—for example, one pair might have the rounded toe and high heel that the customer is looking for but does not come in the right width or has a rubber instead of a leather sole. Before opening the Custom Foot in

Westport, Connecticut, in March 1996, founder Jeffrey Silverman realized that only a collaborative approach could address this customer sacrifice gap.

As in a traditional shoe store, the Custom Foot customer examines physical samples to determine which style he or she desires—but there the similarity ends. Instead of the usual process of having people try on several pairs to find one that fits adequately enough, results from a digital foot imager, measurements taken by hand, and one-to-one conversations with each customer yield a guaranteed fit for each foot. A salesperson then helps the customer choose from a few select design elements to determine the final specifications for the pair of shoes, which are custom-made primarily in Italy.

Mass producers frequently add new features that seek to improve the functionality of existing offerings, such as more types of fasteners on fabric, additional locks and latches on windows, and more gauges and gadgets on manufacturing equipment. Such features generally provide increased value to individual customers, but in many instances they are not enough. Companies—or, worse, customers themselves—are forced to modify the product: clothing is tailored, shoes have pads inserted, windows are remolded, and equipment is realigned. Collaborative customization replaces such back-end solutions with front-end specifications.

It is not surprising, then, that most collaborative customizers focus on design. The design stage, however, is not the only place in the value chain where companies can apply this approach. In the case of collaborative delivery services, customers specify exactly where, when, and how to place goods, which then drives the entire flow of work processes. The personalized placement of meals and groceries by such shopping services as Peapod of Evanston, Illinois, and Takeout Taxi of Herndon, Virginia, is a thriving business today. Unlike mass distributors, which attempt to optimize product supply by forcing customers to come to them, these collaborative customizers not only deliver the product to the customer but also customize that delivery. In effect, there is no supply chain anymore; instead, a *demand chain* is created.

Mass producers scatter a product among as many outlets as possible in the hope that enough customers at enough locations will find the product sufficiently acceptable for it to generate a profit. Inventory is built in anticipation of potential, yet uncertain, demand. Forecasting becomes the critical activity; but, as everyone knows, even the best forecasting models fall short. Even if most companies can accurately forecast their total finished-goods-inventory requirements, they always err in their projections of exactly which goods will be needed at which locations and at what times. Collaborative customizers, in contrast, minimize costs by not keeping inventories of finished products. Instead, they stock raw materials or component parts and then make finished products only in response to the actual needs of individual customers. They transport a given product only to those places where they know it is needed.

ADAPTIVE CUSTOMIZATION

Rather than provide customized offerings, adaptive customizers create standard goods or services that can easily be tailored, modified, or reconfigured to suit each customer's needs without any direct interaction with the company. Each customer independently derives his or her own value from the product because the company has designed multiple permutations into a standard, but customizable, offering. It is the product itself, rather than the provider, that interacts with customers.

Sometimes the technology permits each user to adapt the product—such as the control panels and embedded microprocessors in Lutron's products that enable customers to create different lighting settings. In other instances, however, the technology automatically adapts the product for individual customers. When so-called fuzzy logic or other sensory agents are built into such products as razors, washing machines, and software applications, the embedded technology plays the active role.

If the intrinsic uniqueness of each customer's demands spans an enormous set of possibilities, some form of adaptive customization is imperative. Take the customers that Lutron serves. With the exception of cookie-cutter buildings such as franchise restaurants, every customer's environment is unremittingly different. Each room's shape, decor, and window placement vary. In addition, weather conditions that affect external light change from day to day and hour to hour, as does the composition of people in the room and the way those people use it. Companies that make adaptive mattresses, car seats, and stereo equipment similarly accommodate diverse users wanting to experience the offering differently at different times.

Adaptive customization is the approach of choice also when users want to reduce or eliminate the number of times they have to experiment with all the possible configurations to get the product to perform as they desire. After users of Lutron's Grafik Eye System have made the effort to program a variety of lighting settings, they can select any one of them quickly and effortlessly at any time. Similarly, Peapod has eliminated the "sort-through" sacrifice inherent in going to a physical grocery store filled with more than 30,000 products. Its PC software and on-line service allow customers first to store the personal shopping lists they use to select their purchases and then to access product information through various sorting methods (such as by price, brand, or nutritional content).

Collaboration is the right approach when each customer has to choose from a vast number of elements or components to get the desired functionality or design. But when the possible combinations can be built into the product, adaptive customization becomes a promising alternative for efficiently making many different options available to each customer. For example, Select Comfort of Minneapolis, Minnesota, designs and manufactures mattresses with air chamber systems that automatically contour to the bodies of those who lie on them. Users can select the level of firmness they desire, and couples can select different levels on each side of the bed.

In most cases, adaptive customizers transfer to customers the power to design, produce, and deliver the final goods or services. Electronic kiosks that permit customers to produce their own sheet music, labels, business cards, greeting cards, and other printed materials on the spot illustrate how adaptive customization can put the power to design and manufacture the product directly into individual customers' hands. Similarly, America Online gives its subscribers the ability to create their own stock portfolios that list only the particular equities and funds they own or wish to track. In addition, it offers them a service that automatically delivers articles from various financial publications on the investments in their portfolios, saving them considerable time as well as newsprint-stained hands.

COSMETIC CUSTOMIZATION

A company should adopt the cosmetic approach when its standard product satisfies almost every customer and only the product's form needs to be customized. In doing so, the company visibly demonstrates that it understands the unique ways in which each customer likes the standard product to be presented. In some cases, companies can easily tailor their processes to include simple information about the customer—as simple, in fact, as his or her name—without the dialogue associated with collaborative customization.

Planters knows from daily interaction with its customers that the merchandising philosophies of a warehouse-club store or a convenience store differ from that of a supermarket. It knows that different retail chains' stores allocate a different amount of shelf space to a given product and locate that product differently on the shelf. It even knows which stores plan to feature peanuts on end-aisle displays on a particular weekend. Planters used these insights to develop a customized packaging capability—one that allows each retailer to order the particular product it wants to stock.

Planters understood that its customers had been sacrificing how they wanted to receive and display merchandise. Accordingly, it carefully identified the range of the retail chains' different packaging requirements and then installed new packaging lines that could tailor the package's size, promotional information, and other nonproduct features such as the number of cans wrapped in cartons.

Like Planters, most cosmetic customizers focus their efforts at or near the end of the value chain. Hertz Corporation's #1 Club Gold Program effectively uses cosmetic customization to increase the value of its otherwise standard rental cars. After signing up for the service, Gold Program customers still receive the same basic vehicle, but they bypass the line at the counter and are taken by shuttle bus to a canopied area where they see their own name in lights on a large screen that directs them to the exact location of their car. When customers arrive at the stall, the car's trunk is open for luggage, their name is displayed on the personal agreement hanging from the mirror, and, when the weather demands it (and local laws permit it), the car's engine is running with the heater or air conditioner turned on.

In creating its Gold service, Hertz excelled at identifying which of its existing processes it did not have to change, which new processes it had to add, and which existing processes it could eliminate. It changed reservations, car preparation, and returns. It added the processes for identifying Gold Program customers as they get on the bus, assigning vehicles while customers are en route, and preparing rental agreements automatically. And it eliminated extraneous counter interaction and the time-consuming processes that provided customers with instructions about their car's location. By doing only and exactly what each customer required, Hertz discovered that its Gold service was actually less costly to provide than its standard service.

When performed well, cosmetic customization replaces piecemeal and inefficient responses to customers' requests with a cost-effective capability to offer every customer the exact form of the standard product he or she wants. Both Hertz and Planters were careful not to add processes willy-nilly, which would have resulted in unnecessary complexity and costs. The same cannot be said of the way many mass producers have responded to fragmenting markets. For example, in response to warehouse-club stores' demand for, say, packages containing larger quantities of cereal or additional cans of tuna fish, more than one consumer goods manufacturer today ships cases of products to third-party companies, which in turn take the products out of the cases, shrink-wrap the items in the quantities desired by the club stores, repack the items in a case, and finally ship them on to the stores. The fact that cosmetic customization is easy to pursue does not mean that everyone implements it efficiently.

TRANSPARENT CUSTOMIZATION

Transparent customizers fulfill the needs of individual customers in an indiscernable way—changing the product for them but in such a way that they may not even know that the product has been customized. Instead of requiring customers to take the time to describe their needs, transparent customizers observe behaviors over time, looking for predictable preferences. Of course, this attribute requires a business to have the luxury of time to deepen its knowledge of customers and to move progressively closer to meeting individual preferences. To become a transparent customizer, a business also must have a standard package into which its product's customized features or components can be placed. Transparent customization is the precise opposite of cosmetic customization, with its standard content and customized package.

Businesses ripe for transparent customization are those whose customers do not want to be bothered with direct collaboration. For example, to avoid annoying customers with an endless barrage of surveys on preferences, Ritz-Carlton established a less intrusive means of learning about individual needs. It observes the preferences that individual guests manifest during each stay—preferences for, say, hypoallergenic pillows, classical radio stations, or chocolate chip cookies. The company then stores that information in a database and uses it to tailor the service that each customer receives on his or her next visit. The more someone stays in Ritz-Carlton hotels, the more the company learns, and the more customized goods and services it fits into the standard Ritz-Carlton room—increasing the guest's preference for that hotel over others.

ChemStation likewise gathers information about its customers without their direct collaboration. George Homan, president of Chem-Station, originally defined his business proposition as eliminating a form of environmental waste: the 55-gallon drums that were used to deliver industrial soap and then were discarded in local landfills. After installing ChemStation tanks at numerous customer sites, however, Homan discovered that the real benefit to those customers was eliminating their concern about a necessary but peripheral aspect of their businesses: choosing the proper soap and managing its supplies.

Each customer's purchasing agent, of course, is told that Chem-Station's chemists adjust such factors as pH level, enzyme concentration, foaminess, color, and odor to match the customer's particular needs. But ChemStation determines those needs through its own analysis rather than through collaboration with the customer, and ChemStation alone determines the scheduled frequency of delivery. Soap users develop their own particular habits. For example, customers' employees often use more soap than is necessary—adding that extra glob seems to be a universal habit. Rather than struggle to educate every user about the proper quantity to use, ChemStation may install equipment that regulates the flow of active ingredients or that dilutes the mixture of detergent and water by the appropriate amount (while, of course, giving the purchasing agent the reason for the concomitant price reduction). Only ChemStation knows the precise formula each customer uses and the reasons for its selection, which has the added benefit of keeping customers from educating the competition.

The ChemStation tank is the standard package that contains the customized goods (the soap) and service (the delivery of the soap). Customers never think about the soap *getting* there, just about its always *being* there. By constantly monitoring inventory levels in its tanks, ChemStation can learn how often customers will need more soap and can deliver it so that they always have the proper amount, saving them the bother of having to monitor supplies and place orders. Because there is no need to notify customers when deliveries are to be made or even that they have been made, ChemStation is able to construct very cost-effective delivery routes. The customer simply reviews its usage and pays the invoice at the end of each billing period.

Combining Multiple Approaches

Each of the four customization approaches used alone challenges the mass production paradigm of offering standard goods or services to all customers. Many companies, however, combine two or more approaches. For example, Lutron, predominantly an adaptive customizer, nonetheless collaborates with customers to match the color of its products to their walls or to integrate its lighting controls with their security systems. Similarly, Planters, primarily a cosmetic customizer, periodically collaborates with retailers to change the mix of nuts they receive.

The key is to draw on *whatever* means of customization prove necessary to create customer-unique value. Datavision Technologies Corporation, a San Francisco producer of marketing materials, effectively combines three of the approaches: collaborative, cosmetic, and transparent. The company takes input from multiple sources to mass-customize materials for marketing such products as financial plans, vacation packages, corporate health care programs, and cars. It draws from a vast library of materials in order to produce videos that are coupled with print information to create messages tailored to individuals' specific needs.

Datavision produces the customized videotapes with a computer-controlled process that employs laser disc players, graphics generators, and video recorders. A detailed profile of each customer's interests and past purchase history drives the process. The system links each element of the customer's profile with specific video, voice-over, music, graphics, and other text segments. It then automatically assembles the script and presentation modules. Each videotape is assigned an identification number that is used to print customized packaging materials, including cassette labels, mailing labels, and other printed materials accompanying the videotape. The process can mass-customize individual videotapes in small quantities as well as in batches of tens of thousands.

Datavision has produced several marketing programs for automobile manufacturers. Whenever customers call a client's toll-free number for information on a specific car model, the telemarketing employee works with them to identify the car attributes they find most important and to learn what competing models they also are considering. This interaction carries over to the videotape that customers receive. The opening segment provides a checklist of the specific car attributes mentioned in the telephone conversation, complete with a voice-over reminding customers of their stated remarks. If Jane Jones mentions an advanced engine as an important attribute, then her video might include a computerized graphic of the engine with a high-tech music track and a voice-over on the engine. If Robert Smith regards the power train as an important attribute, his video might include a sports music track and information on the car's horsepower and torque.

Datavision's interaction with customers to identify the aspects of the product that matter most to them is collaborative customization. The selection of the video clips and their sequencing, the voice-over, and the music—all of which are based on what the company can easily glean from each conversation—are transparent customization. Datavision uses cosmetic customization when the customer's name appears on the tape's label and in the opening titles: "This video presentation produced especially for Jane Jones." The customer's name is not spoken to avoid making mistakes in pronunciation; but as the name appears on the screen, the narrator says the tape was made "for you," "for you and your husband," or "for you and your wife," depending on the information the customer provided. It is the combination of the three approaches that produces an effective and relevant marketing message.

The four approaches to customization provide a framework for companies to design customized products and supporting business processes. They demonstrate the need to mix the direct interaction of collaborative customization, the embedded capabilities of adaptive customization, the forthright acknowledgment of cosmetic customization, and the careful observation of transparent customization into one's economic offerings. Customers do not value merchants who recite monolithic mantras on customer service; they value—and buy—goods and services that meet their particular set of needs. There is a time to conduct a dialogue with customers and a time to observe silently, a time to display uniqueness and a time to embed it. Businesses must design and build a peerless set of customization capabilities that meet the singular needs of individual customers.

8
Versioning: The Smart Way to Sell Information

Carl Shapiro and Hal R. Varian

In 1986, Nynex issued the first electronic phone book, a compact disc containing all the telephone listings for the New York area. Charging $10,000 a copy, the company sold the CDs to the FBI, the IRS, and other large commercial and governmental organizations. Sensing a great business opportunity, the Nynex executive in charge of the project, James Bryant, left to set up his own company, Pro CD. His goal was to produce an electronic directory covering the entire United States.

The phone companies, fearing an attack on their lucrative yellow pages businesses, refused to license digital copies of their listings to Pro CD. But that didn't stop Bryant. He went to Beijing and hired Chinese workers—at $3.50 a day—to type into computers every listing from every U.S. telephone book. The resulting database, containing more than 70 million phone numbers, was used to create a master disc, which in turn was used to create hundreds of thousands of copies. The CDs, which cost well under a dollar each to produce, sold for hundreds of dollars, yielding a tidy profit for Pro CD.

But the CD–phone book boom was short-lived. Attracted by the seemingly strong profit potential, competitors such as Digital Directory Assistance and American Business Information rushed to launch competing products containing essentially the same information. Because their products were indistinguishable, the companies were forced to compete on price alone. Not surprisingly, prices plunged. Soon, CD phone directories were selling for a few dollars in discount software bins. A high-priced, high-margin product just months before, the CD phone book had become a cheap commodity.

The rapid rise, and even more rapid fall, of CD telephone directories stands as a cautionary tale for the purveyors of information products, particularly those sold in digital form. It reveals that the so-called new economy is still subject to the old laws of economics. In a free market, once several companies have sunk the costs necessary to create an undifferentiated product, competitive forces will usually move the product's price toward its marginal cost—the cost of manufacturing an additional copy. And because the marginal cost of reproducing information tends to be very low, the price of an information product, if left to the marketplace, will tend to be low as well. What makes information products economically attractive—their low reproduction cost—also makes them economically dangerous.

Many information producers make the mistake of assuming that their products are exempt from the economic laws that govern more tangible goods. But, as Pro CD found out, that's just not so. Although information goods have unusual production economics, they are nevertheless subject to the same market and competitive forces that govern the fate of any product. And their success, too, hinges on traditional product-management skills: gaining a clear understanding of customer needs, achieving genuine differentiation, and developing and executing an astute positioning and pricing strategy.

Information's Dangerous Economics

To forge a winning strategy for an information product, you need to understand the economics of information production. Information goods, which we define as goods capable of being distributed in digital form, have always been characterized by a distinctive cost structure: producing the first copy is often very expensive, but producing subsequent copies is very cheap. A book publisher, for example, may spend hundreds of thousands of dollars to acquire, edit, and design a manuscript, but once the first copy of the book has been printed, the cost of printing another is usually only a few dollars. To get a movie made, a producer may spend a hundred million dollars on cast, crew, script, and sets, but making a print of the final cut will cost only a few hundred dollars. The fixed costs of producing information are large, in other words, but the variable costs of reproducing it are small.

The sharp skew toward fixed costs is not the only thing distinctive about the cost structure of information goods. The fixed costs and the variable costs themselves have unusual characteristics. The fixed costs tend to be dominated by sunk costs—costs that are not recoverable if production is halted. If you invest in a new office building or factory and later decide you don't need it, you can recover part of your fixed costs by selling the facility. But if your film flops, you probably won't be able to sell off the script or the sets, and if your CD is a dud, it ends up in the cut-out racks at $4.95.

The variable costs of producing information also have a unique feature: the unit cost of creating an additional copy of an information product typically does not increase even if a great many copies are made. Information producers, in other words, have few capacity constraints, which is quite a different situation from that faced by most manufacturers. If sales of microchips grow, for example, Intel will at some point need to build an expensive new fabrication facility to meet the added demand. And if sales of airplanes increase, Boeing will have to invest heavily in new plants, machinery, and people. When these and other traditional manufacturers reach the limit of their existing capacity, the cost of producing an additional unit goes way up. That doesn't happen with most information products, which can be reproduced with a high degree of automation at very low cost. If you can make one copy, you can make a million copies, or ten million copies, at roughly the same unit cost.

Because of their cost structure, information products offer vast economies of scale: the more you produce, the lower your average cost of production. That's why Microsoft, with its dominance in personal-computer operating systems and business applications, enjoys gross profit margins of 92%. But the cost structure has a big downside as well. Because the fixed costs are both large and sunk, companies that don't enjoy market dominance can be caught in devastating price wars. If competition forces a company to reduce its prices to a level near its marginal production costs—as was the case with publishers of CD phone books—that company will never be able to recoup its big up-front investments. It will, in time, face economic doom. What economists call a perfectly competitive market represents a disaster scenario for information producers.

The dangers inherent in the economics of information production become even more pronounced when the information is produced digitally. Copies of information in digital form—such as CDs or digital video disks—are much cheaper to reproduce than analog or print copies. Think of encyclopedias. Printing an encyclopedia set can cost more than a hundred dollars. Cutting a CD-ROM of that same set costs just pennies. By reducing variable costs, digital reproduction further exaggerates the skew toward fixed costs.

And that's not all. When digital information is delivered over a network, the variable costs can disappear almost completely. Because the product has no physical form—it exists purely as bits of data—there's no cost for manufacturing, no cost for packaging, no cost for shipping. Once the first copy of the information has been produced, transmitting additional copies is essentially free. Consider again the case of electronic phone books. Today, if you want to quickly look up phone numbers for people around the country, you don't even need to buy a CD—no matter how cheap they've become. You can search phone listings for free at dozens of Web sites. It costs next to nothing to let an additional customer search an on-line database, so competing providers give the information away to all comers, hoping to make their money by selling ads.

Many commentators have marveled at the amount of free information on the Internet, but to economists like us it's no surprise. The generic information flowing through cyberspace—phone numbers, news stories, stock prices, maps, and the like—is simply selling at its marginal cost: zero. (See "The Logic of the Free Version.")

The Logic of the Free Version

A refrigerator manufacturer would quickly go broke if it started handing out free samples of its products. But information producers give away free versions all the time. When you buy a computer, you are likely to find a bundle of free software on the hard drive. When you surf the Web, you discover not only oceans of freely distributed news stories, statistics, and other information, but also loads of software demos and "freeware" that you can download with a few clicks of your mouse.

Free versions of digital goods are common for two reasons. First, the low marginal cost of creating copies of information means that it doesn't require much, if any, investment to give information away. Second, information is an "experience good"—customers don't know what it's worth until they've actually tried it. Free versions provide customers with an easy and attractive way to test out a digital product.

Of course, if all you did was give information away for free, you wouldn't have much of a business. (Many Web site operators are learning this lesson the hard way.) There has to be sound business logic behind the free offer. We have found that the most savvy information producers offer free versions only when they are likely to achieve one or more of the following goals:

Building Awareness. Many companies use free versions simply to create awareness of their products. They give away a version with limited content or features in order to entice consumers to pay for the full version. Computer game makers, for example, distribute free demos over the Web or on CD-ROMs bundled with game magazines. The demos allow users to play only the first few levels of a game—enough to get hooked but not enough to get bored. The hope is, of course, that the user will rush out to buy the complete game. Using a free version to build awareness works for games because each game is unique; the customer can't buy a substitute. But if you're only one of many providers of the same information, a free version may simply underscore the commodity nature of your product. The customer, spoiled by the free version, will be motivated to find the cheapest possible source of the information.

Gaining Follow-on Sales. A more sophisticated strategy is to give away a version in order to build a base of customers to which you can sell follow-on products, such as extensions, upgrades, and services. The free version, in this case, is usually a complete version; no content is missing and no features are disabled. The idea is to get customers to become dependent on the product. The more they use it, the more interested they'll be in add-ons. McAfee Associates, for example, offers many of its core virus-protection products for free over the Web. It makes money by selling site licenses to companies, upgrades to individuals, and an array of services to both groups. McAfee's products now account for half the sales of antivirus software.

Creating a Network. Because many digital goods are subject to network effects—they only become valuable once a large number of people are using them—free versions can also be a good way to bring a product's use up to a critical mass. Adobe, for example, gives away a simple version of its Acrobat software that enables users to view and print electronic documents even if they lack the software the documents were created with. Because Adobe was the first to seed the market with such a program, it is now able to sell full versions of Acrobat—at between $200 and $600—to anyone who wants to share electronic documents over the Web or through other media.

Attracting Eyeballs. As the war for attention continues to intensify on the Internet, free information becomes ever more valuable as a lure to attract the eyes of surfers. Some information companies, in fact, are finding that they can earn more money from advertising than from selling their own information. Playboy, for example, posts free images of its playmates on its Web site, along with banner advertisements that it sells for more than $10,000 a month. Each image incorporates a digital watermark, enabling Playboy to track not only how many people view the image on its own site but also how many people view it after it's been copied onto other sites. In this way, Playboy learns more about its on-line customers and how they use its products, further strengthening its ability to sell ads as well as on-line and print subscriptions.

Gaining Competitive Advantage. Sometimes the strategic value of getting a large number of people to use your information is greater than the economic value of getting a smaller number of people to buy it. Microsoft, for example, gives away its Internet browser in order to prevent Netscape from gaining control over computer desktops. That competitive benefit far outweighs the money it could make by selling the browser. Microsoft sometimes finds itself on the other end of such strategies, too. One of the main reasons Sun Microsystems gives away many of its Java programming tools is to reduce the market power wielded by Microsoft. Because Java can be used with any computer, it makes operating systems like Windows 98 relatively less valuable and hardware, such as Sun's servers, relatively more valuable.

Linking Price to Value

The extremely low marginal costs of information production rule out many traditional pricing strategies. You can't, for example, use cost-based pricing. Nor can you set prices according to the competition—that's a sure road to ruin. The only viable strategy is to set prices according to the value a customer places on the information.

But which customer? The value of a piece of information can vary dramatically from one person to the next. A stock market speculator will place a far greater value on stock quotes than will a long-term investor who buys and holds. A computer "power user" will value the latest operating-system upgrade much more than the average home user will. And a drug company executive will likely place more value on the text of the latest FDA rulings than a pharmacist, who, in turn, will place greater value on it than a premed student. Information never has the same value for every potential customer.

In a perfect world, an information producer would sell its product to each buyer at a different price, reflecting the value that the different buyers place on it. In reality, though, such personalized pricing is rarely possible. For one thing, even in these days of cheap computing, it is awfully expensive to capture, store, and distribute data on the tastes of individual customers. For another, traditional sales channels, like retail stores, cannot set an array of prices for the same good. (Even if they could, it would be next to impossible to get customers to stay within their intended pricing strata—just look at all the gyrations airline customers go through to locate the cheapest routes.) And finally, information producers run the risk of annoying or even alienating their customers if they charge different prices for the same product.

But there is a practical way to set different prices for basically the same information without incurring high costs or offending customers. You do it by offering the information in different versions designed to appeal to different types of customers. With this strategy, which we call *versioning*, customers in effect segment themselves. The version they choose reveals the value they place on the information and the price they're willing to pay for it.

Traditional information providers have always used versioning, in one form or another, as a way to structure their product lines. Publishers release a book first in hardback and later in paperback, selling the same text at a high price to readers who must have the book right away and at a lower price to people who don't mind waiting. In a similar way, movie houses charge $7 or more for a ticket to a film that can be rented six months later for $3 a household.

When information is produced digitally, versioning becomes an even more flexible and powerful strategy. For one thing, it's easy to manipulate digital data, so the cost and time required to produce and distribute different versions go way down. For another, the proliferation of CD-ROM players, VCRs, and Internet browsers opens many kinds of information to a much larger and more diverse audience. When legal information was conveyed only in heavy and expensive tomes, lawyers were the only people interested in purchasing it. Now that the information can be searched and bought by the bit, there are many more potential customers for it. Versioning provides a way to sell information to those customers in a form that they will value without cannibalizing the existing high-price, high-margin market.

The trick is to identify the best ways to distinguish the different versions of your product. You need to determine which features will be highly valuable to some customers but of little value to others. Then you need to create the right number of versions and set the right prices for them. The goal is to get each customer to pay the highest possible price for the product, thus maximizing the overall returns. Since the customers themselves are selecting the price they'll pay, based on their own calculation of the information's value, they will be far less likely to take offense at paying different prices than they would if the manufacturer were imposing the prices on them.

The Many Versions of Versioning

In the past, versions of information products were usually based on timing or, more precisely, delay. For almost any type of information, some people will always be more eager to get their hands on it than will others. That's the rationale for releasing hardcovers before paperbacks and for showing movies in theaters before putting them on tape. Delay is often a good basis for versions of digital information as well. PAWWS Financial Network, for example, offers two versions of its portfolio accounting system, one at $8.95 a month and the other at $50.00. What's the difference? The inexpensive service uses stock quotes that are delayed by 20 minutes to calculate portfolio values, whereas the premium service uses real-time quotes. Those 20 minutes are very valuable to one set of the company's customers.

But with digital information, delay is only one of many possible dimensions for versioning. Just consider the wide variety of ways in which digital products are differentiated today:

Convenience. Restricting the time or place at which a customer can access information, or restricting the length of access, is often a good way to get buyers to reveal the value they place on the information. The more a customer needs the information, the more freedom they'll want in accessing it. America Online, for example, offers different monthly membership plans based on convenience. The standard plan, which provides unlimited access, costs $21.95. An alternative plan costs $4.95 but allows only three hours of connection time—if you use more, you pay a high hourly surcharge. By offering the cheaper version, AOL can attract customers who have only a limited need for its service—they may use it solely for e-mail, for example—while maintaining much higher prices for customers with a greater dependence on it. Similarly, some on-line database companies offer discount subscriptions to users who agree to log on only outside of normal business hours.

Comprehensiveness. Some customers will pay a big premium for information that offers a depth of detail—in geographical coverage, historical scope, or statistical detail. Public affairs specialists and journalists, for example, will value the ability to search the full text of articles from newspapers around the world. Many scholars and students will value extensive historical information. Marketing managers will value information on individual customers and their long-term purchasing patterns.

Many newspapers and magazines are using comprehensiveness as the basis for creating versions of their on-line products. The *New York Times* and *Business Week,* for example, give away their current editions' content on the Web, but they sell access to their extensive archives. Because there are so many sources of news on the Internet, these publications know that the only way to attract readers—and in turn advertisers—is to give away their freshest content. But they can charge for their past articles because the segment of customers that values those articles—writers, researchers, and the like—have no other practical source for them.

Manipulation. Another important dimension that can form the basis for versioning is the ability of the user to store, duplicate, print, or otherwise manipulate the information. Back in the days of copy-protected software, companies like Borland sold two versions of their programs—one was low priced and could not be copied and the other was high priced and could. Many information providers today use similar constraints on information manipulation to distinguish their products. Lexis-Nexis, for instance, imposes additional charges on users who want to print or download information rather than just view it on screen.

Community. The chat rooms and bulletin boards that crowd the Web demonstrate that many people value the opportunity to discuss information with others who have similar interests. By restricting users' ability to join an on-line community, providers can identify customers who place value on the community in addition to the information.

Silicon Investor, a popular Web site for investors in high-tech industries, offers hundreds of discussion boards on individual companies. It allows anyone to read the messages posted on the boards for free. But if you want to post a message—or send a private e-mail to another member—you have to pay an annual membership fee of $100 or a lifetime fee of $200. By allowing free access to the information on the site, Silicon Investor gets more people to visit the site, enabling it to charge more for advertising. By charging an extra fee for posting messages, it makes money on customers who want to do more than read.

Annoyance. People who pay the Silicon Investor membership fee also get an added benefit: they have the capability to turn off the advertisements posted throughout the site. The ability to avoid the annoyance of on-line ads is valued by some Web surfers, and they're willing to pay extra for it.

Similarly, many shareware programs, which are distributed free for trial use, incorporate a start-up screen that asks users if they're ready to purchase a registered version. The only way to avoid the annoying screen is to send in money.

Speed. A common strategy for software makers is to sell versions of their programs that run at different speeds. The most serious users naturally gravitate to the faster versions even if they have to pay a lot more for them; the greater efficiency outweighs the higher cost. Wolfram Research, for example, used to sell two versions of Mathematica, its program for performing symbolic, graphical, and numerical mathematics. The high-priced professional version used a computer's floating-point processor to speed up the calculations. The cheaper student version disabled the processor, slowing the calculations considerably.

Interestingly, Wolfram had to write more code to get the student version to work without the floating-point processor. The inexpensive version thus cost more to produce than did the premium version. But offering the low-speed version made economic sense because it expanded the overall user network, making the professional product even more valuable to the sophisticated users, such as professors who wanted to share files with their students. (See "Value-Subtracted Versions.")

Value-Subtracted Versions

A few years back, IBM offered two very similar versions of one of its printers—the high-end LaserPrinter and the less expensive LaserPrinter E. The two versions looked the same and functioned the same, with one exception: the LaserPrinter could print ten pages per minute while the E version could print only five. A testing lab for computer equipment examined the two models and found that a special chip had been inserted into the LaserPrinter E to slow down its operation. IBM had, in other words, deliberately degraded the performance of its high-end model in order to create a cheaper model. And because the subtraction of value required the manufacture and installation of a special chip, the low-priced version actually cost more to produce than the high-priced one.

IBM's tactic was unusual. When most manufacturers want to create versions of their products, they start by building a bare-bones model, then add features to create premium versions. The high-end models cost more to create than the low-end alternatives. Toyota, for example, spends considerably more to produce a top-of-the-line Camry XLE, with leather upholstery, antilock brakes, and traction control, than it does to manufacture an entry-level Camry CE that lacks the luxury features.

For digital goods, however, the IBM method is the rule, not the exception. Most versions of digital information are created by subtracting value rather than by adding it. The producer first invests in developing the most technologically advanced version—in order to have a distinctive product that will appeal to the most demanding and least price-sensitive customers—and then removes features or capabilities to tailor the product to less demanding customers. As the low-end users' needs advance, they can follow an established upgrade path within the same product family.

PhotoDisk, for example, scans its stock photos at high resolution to create its premium product, then degrades them to produce low-resolution copies. Offering the low-resolution versions requires extra work—and extra server space for storage—so it actually costs the company more to produce its cheap product than its expensive one.

Charging less for products that cost you more to produce may sound illogical, but for digital goods it makes sense. The extra investment required to create degraded versions is usually modest and can be recouped quickly as sales grow. And the revenue from versions designed to appeal to different market segments helps offset the big fixed costs required to create the product initially.

There's an important caveat to value subtraction: you have to make sure that customers can't transform the degraded version back into the original. With a world full of talented hackers, that's no easy feat. There have been reports, for example, that users of Microsoft's $250 workstation version of its NT software have figured out how to turn it into the original and more powerful $1,000 server version with just a few simple tweaks of the code. When you choose the dimensions of your product to manipulate to create different versions, choose carefully.

Data Processing. Various data-processing capabilities can often be built into an information product, enabling certain users to carry out sophisticated tasks. H&R Block, for example, offers the standard version of its Kiplinger's TaxCut software to people who just want an automated way to fill in their tax forms. But it also offers a pricier premium version, TaxCut Deluxe, that includes a number of other tools—for example, it has an audit feature that examines your return and highlights entries likely to catch the attention of IRS agents.

User Interface. Varying the way that customers access information can be a particularly good basis for versioning. Sophisticated users will often be willing to invest time learning a complex interface that offers, for example, powerful searching capabilities. (And their up-front investment of time will make them less likely to shift to a competing product later.) More casual users will want a simpler, more intuitive interface even if its capabilities are rudimentary. Adobe's $600 Photoshop software for manipulating photographic images has a complex interface intended for professional designers. But the company also sells a lower-end product, the $50 PhotoDeluxe, that has a stripped-down interface geared for home users. You can't do as much with PhotoDeluxe, but you don't have to spend a lot of time learning how to use it, either.

Image Resolution. Many digital products include images, and different users will place different values on the quality of the images. The stock-photo house PhotoDisk, for example, offers its photographs over the Web at different resolutions. Professional designers creating glossy brochures purchase high-resolution images at $49.95 each, but newsletter producers settle for lower resolution images at $19.95. The myriad pornography sites on the Web often offer low-quality "thumbnails" of their photographs for free but charge for the ability to view and download high-resolution copies.

Support. Some information providers offer different levels of technical support at different prices. You can, for example, download Netscape's Web browser for free over the Net or, for $40, you can become a subscriber and receive not only the software but also an instruction manual and one free phone call to a support technician. Using technical support as a basis for versioning can be tricky, though. Highlighting the added value of support may raise questions in customers' minds about the reliability of the product. And failing to deliver on promises of support can turn into a public relations nightmare.

In addition to being used in isolation, the different dimensions of versioning can also be combined. Dialog, the large on-line information provider, creates versions of its Web-accessible database by altering both its user interface and its comprehensiveness. A high-end version, DialogWeb, is designed for corporate researchers and other information professionals. It has a powerful but complex interface, allowing highly sophisticated searches, and offers access to the full range of Dialog's content. Another product, DataStar, is much cheaper and much less powerful, offering a subset of the full Dialog database with a simplified interface. DataStar suits casual users well, but because of its limitations it does not siphon away professional users from DialogWeb.

The Mechanics of Versioning

So how many versions should you offer? There's no pat answer to that question. The number should be guided by two considerations: the characteristics of the information that you're selling and the value that different customers place on it. If your information can be used in many ways, it probably makes sense to offer a wide array of versions. But if the value of your information hinges on the number of users who access it in the same format—if, in other words, the information is subject to network effects—you may want to restrict the number of versions you offer.

Kurzweil Applied Intelligence, for example, offers many versions of its voice recognition software. Kurzweil understands that voice recognition has many different applications and that they vary greatly in the value they provide to users. College students will be attracted to a simple product that enables them to create documents by speaking into their computers. But given their limited budgets, they'll only buy such a product if its price is low. Doctors, on the other hand, will be drawn to a highly sophisticated product that is able to understand a specialized vocabulary—and because such a product will save them a lot of time, they'll pay handsomely for it.

To capture the different levels of customer value, Kurzweil offers seven different versions of its software, distinguished mainly by the

Exhibit 8-1   One Product, Many Versions

Recognizing that its voice recognition software can be used in many ways, Kurzweil Applied Intelligence offers a broad array of versions at very different prices.

Version	Price	Vocabulary
VoicePad Pro	$79	20,000-word general
Personal	$295	30,000-word general
Professional	$595	50,000-word general
Office Talk	$795	business
Law Talk	$1,195	legal
Voice Med	$6,000	medical
Voice Ortho	$8,000	special-purpose medical

size and specialization of the vocabularies they recognize. The top-of-the-line, $8,000 version for surgeons, Voice Ortho, is 100 times more expensive than the $79 entry-level product for students, VoicePad Pro. Between those extremes are versions tailored to home users, business users, and lawyers, all at different price points. Because each segment's needs are unique, there's little chance that buyers will be confused by the various options. And there's also little chance that customers targeted for high-priced versions will opt instead for lower-priced versions. A version unable to recognize legal terms, for example, would have little value for an attorney. The way customers define the value of the product locks them into their intended segment. (See Exhibit 8-1 "One Product, Many Versions.")

For other companies, a more limited array of options makes sense. Intuit, for example, offers only two versions of its popular Quicken software for personal financial management. Unlike Kurzweil's product, Quicken doesn't have a wide variety of applications—a lawyer balances her checkbook in pretty much the same way a doctor does—so having to choose from a broad range of versions would simply confuse customers. By limiting the number of versions, Intuit gets other benefits as well. Customer support stays simple, and users are able to share files with less risk of incompatibility.

But by offering only one high-end product and one low-end version, Intuit, like many information companies, may be missing out on an important opportunity. The two-version strategy, though enticing in its simplicity, ignores the psychological phenomenon known as "extremeness aversion." When buying products, consumers normally try to avoid extreme choices—they fear they'll pay too much if they go for the most expensive version, and they worry they'll get too little if they opt for the cheapest. They are drawn instead to a compromise choice—a version in the middle of the product line. Like Goldilocks, they don't want "too big" or "too small"—they want the product that's "just right."

By offering three versions of a product, companies can shift buyers away from the entry-level product and to the more expensive middle offering. The effect can be quite dramatic. In one experiment, researchers offered customers different sets of microwave ovens. When the choice was between a no-frills oven at $109.99 and a midrange model at $179.99, customers chose the midrange oven 45% of the time. When a high-end oven at $199.99 was added to the choice set, people chose the midrange oven 60% of the time. The existence of this phenomenon is the reason McDonald's offers its drinks in three sizes rather than just two.

Information producers can also capitalize on extremeness aversion. If they are currently offering only two versions, they should consider adding a third, high-end product to their line. If Intuit, for example, offered a third version of its Quicken software—Quicken Gold, say—at a price higher than that for the Deluxe version, many buyers who would have bought the standard product will instead move up to Quicken Deluxe. The important thing to recognize is that the product you really want to sell should be positioned in the middle—the high-end product is there mainly to pull people toward the compromise choice.

The optimum number of versions to offer is, as we've seen, rarely clear-cut. The best way to decide is often through trial and error. Because it's usually inexpensive to create new versions of an information product, a company can do a lot of experimentation. Recently, for example, Information America, a company that provides public records to banks, government agencies, and law offices, was trying to decide whether to offer its services to users operating at home. The company felt that demand would be high enough to justify entry into the new market, but it was concerned that a price low enough to attract home users might cannibalize its sales to professionals. To gain insight into the problem, the company created a subsidiary, KnowX, to offer home users access to a subset of its databases via the Web. It turned out that the restricted offering was very popular—and it didn't attract the high-end professionals. With today's powerful technologies for distributing information, more and more companies are, like Information America, finding it easy to explore market segments that were not reachable before.

Old Ideas, New Applications

Information has always played a central role in our economy—a simple fact that too often gets lost in all the hype about the information age. And the total amount of information in existence hasn't expanded all that much in recent decades. What has changed is that the information has become dramatically more accessible. Many of the great technological advances of the twentieth century—telephones, radio, motion pictures, television, computers—have served to speed the flow and widen the availability of information. The arrival of the Internet is just the latest step—albeit a very big step—in a process that continues to unfold.

As access to information has expanded, so too have the opportunities for selling information goods to a broader and more diverse set of customers. Versioning provides a way to serve that larger market by tailoring the same core of information to the needs of different buyers. It not only enables you to gain more revenue from an existing product but also provides a basis for thinking creatively about how to distinguish your product from competing offerings. By monitoring how the market reacts to new versions, you gain ever greater insight into how customers define value, allowing you to continually refine your product line. A creative versioning strategy is often the best defense against the commoditization of information.

Success in selling digital goods does not require a whole new way of thinking about business. Rather, it requires the same kind of smart managing and smart marketing that have always set apart the best companies. The real power of versioning is that it enables you to apply tried-and-true product-management techniques—segmentation, differentiation, positioning—in a way that takes into account both the unusual economics of information production and the endless malleability of digital data.

9

Making Mass Customization Work

B. Joseph Pine II, Bart Victor, and Andrew C. Boynton

Continuous improvement at Toyota Motor Company is now a business legend. For three decades, Toyota enlisted its employees in a relentless drive to find faster, more efficient methods to develop and make low-cost, defect-free cars. The results were stupendous. Toyota became the benchmark in the automobile industry for quality and low cost.

The same, however, cannot be said for mass customization, Toyota's latest pioneering effort. With U.S. companies finally catching up, Toyota's top managers set out in the late 1980s to use their highly skilled, flexible work force to make varied and often individually customized products at the low cost of standardized, mass-produced goods. They saw this approach as a more advanced stage of continuous improvement.

As recently as early 1992, Toyota seemed to be well on its way to achieving its goals of lowering its new-product-development time to 18 months, offering customers a wide range of options for each model, and manufacturing and delivering a made-to-order car within three days.

In the last 18 months, however, Toyota has run into trouble and has had to retreat, at least temporarily, from its goal of becoming a

Authors' note: IBM Consulting Group partners and consultants contributed significantly to the development of the ideas in this article. The research was sponsored in part by the IBM Consulting Group and the IBM Advanced Business Institute.

mass customizer. As production costs soared, top managers widened product-development and model life cycles and asked dealers to carry more inventory. After Toyota's investigations revealed that 20% of the product varieties accounted for 80% of the sales, it reduced its range of offerings by one-fifth.

What happened? Was Toyota's new goal off-base in the first place, or was the mass-customization program a victim of troublesome economic times? Many analysts believe that Japan's recession and the devaluation of the dollar against the yen were the culprits that forced Toyota's pullback. These factors had undermined the company's competitive position and were causing its profits to slide. But, according to Toyota top managers, these weren't the only reasons for the company's retrenchment. They acknowledged that they had learned the hard way that mass customization is not simply continuous improvement plus.

All too often, executives at manufacturing as well as service companies that have been pursuing continuous improvement do not realize that mass customization is a distinct and, generally, a very unfamiliar way of doing business. This mistake is understandable. The frequent process enhancements generated by continuous improvement can increase the inherent flexibility of those processes. And, as a work force gets better and better, expanding its range of skills, it can handle an increasingly complex set of tasks, such as assembling a variety of products or delivering tailored services.

While executives are correct in thinking that continuous improvement is a prerequisite for mass customization, one thing is becoming clear from the experiences of companies such as Toyota, Amdahl, and Dow Jones. Continuous improvement and mass customization require very different organizational structures, values, management roles and systems, learning methods, and ways of relating to customers. (See "Understanding the Differences.")

Understanding the Differences

Mass Production
The traditional mass-production company is bureaucratic and hierarchical. Under close supervision, workers repeat narrowly defined, repetitious tasks. Result: low-cost, standard goods and services.

Continuous Improvement

In continuous-improvement settings, empowered, cross-functional teams strive constantly to improve processes. Managers are coaches, cheering on communications and unceasing efforts to improve. Result: low-cost, high-quality, standard goods and services.

Mass Customization

Mass customization calls for flexibility and quick responsiveness. In an ever-changing environment, people, processes, units, and technology reconfigure to give customers exactly what they want. Managers coordinate independent, capable individuals, and an efficient linkage system is crucial. Result: low-cost, high-quality, customized goods and services.

In continuous-improvement systems, tightly linked teams bridge disparate functions that typically interact with each other in a predictable, sequential manner. A hallmark is the conviction that every process must contribute to satisfying the customer by constantly and incrementally achieving higher quality. But a big difference from mass-customizing systems is that workers do not question the basic design of the product that they are assigned to build; they assume it to be what customers want.

Continuous-improvement organizations school workers in tools and techniques to help them improve the tasks they must perform. The fundamental precept is to learn by doing a task and then do it better. Managers of such organizations lead everyone on a relentless mission to eliminate waste and enhance quality through a vision of "being the best," while still ensuring reliable outcomes from routine tasks. These managers are eternally striving to tighten the links between processes so that every team and individual worker knows how its function affects others and ultimately the quality of the product or service. They must be coaches who constantly urge employees to interact, converse, improve, and do what is right for the team. They try to foster values that create a sense of community because the interests of the individual are subsumed within the interests of the team, the company, and the customer.

Mass customization, on the other hand, requires a *dynamic network* of relatively autonomous operating units. Each *module* is typically a specific process or task, like making a given component, a distinctive welding method, or performing a credit check. The modules, which

may include outside suppliers and vendors, typically do not interact or come together in the same sequence every time. Rather, the
combination of how and when they interact to make a product or provide a service is constantly changing in response to what each customer wants and needs. From continually trying to meet these
demands, the mass-customization organization learns what new capabilities it requires. Its employees are on a quest to increase their own
skills, as well as those of the unit and the network, in a never-ending
campaign to expand the number of ways the company can satisfy
customers.

Managers in these ever-changing settings are coordinators whose
success depends on how well they perfect the links that make up the
dynamic network. They strive to make it ever easier and less costly for
the process modules to come together to satisfy unique customer requests. And they lead the effort to increase the range of things that the
organization can do. They must create a culture that places a high
value on the diversity of employees' capabilities because the greater
the diversity of the modules, the greater the range of customization
the organization can offer.

What all this boils down to is that mass customization is a totally different world from continuous improvement. It is a world in
which the unpredictable nature of each customer's demands is considered an opportunity. To exploit that opportunity, the organization must perpetually generate new product teams. The key to success is designing a linkage system that can bring together
whatever modules are necessary—instantly, costlessly, seamlessly, and
frictionlessly.

When Mass Customization Cannot Work

Continuous improvement can certainly be a subset of mass customization. The autonomous operating units within a mass customizer can
and should strive to continuously improve their processes. But as Toyota, for one, seems to have finally realized, mass customization generally cannot be a subset of continuous improvement.

One of the main causes of Toyota's recent problems was that it had
been pursuing mass customization but had retained the structures
and systems of continuous-improvement organizations. By doing this,

Toyota ended up not succeeding at mass customization and, at the same time, undermining its continuous-improvement efforts.

For example, Toyota assumed that its work force had attained the skills needed to handle production of its rapidly growing range of product offerings. But when the frequently changing tasks butted up against the limit of workers' capabilities, managers did not realize that the problems stemmed from a failure to transform the organization. Rather than developing the loose network necessary to make a mass-customization organization work, Toyota managers turned to machines. Over time, this ended up weakening the skills of the workers and thus violated an essential tenet of continuous improvement. It also caused internal friction.

One action Toyota took was to invest heavily in robots. But as one Toyota manager later commented, "Robots don't make suggestions." Toyota also installed monitors at some stations along the assembly line that told workers how to put together a particular car. And the company installed computer-controlled spotlights illuminating the bins containing the right components. These measures deprived employees of opportunities to learn and think about the processes and, therefore, reduced their ability to improve them.

Another big problem at Toyota was that product proliferation took on a life of its own. Like mindless continuous improvers, engineers created technically elegant features regardless of whether customers wanted the additional choices. In mass customization, customer demand drives model varieties.

A third problem arose when Toyota's management, in its pursuit of low-cost customization, pushed product development teams to use more common components across its models. At Toyota, project leaders have overall responsibility for the development of a given model, but separate teams develop individual components, such as brake systems or transmissions, which ideally will be used in several models. Project leaders felt that the intensifying pressure to share components was forcing them to compromise their models, and they began to resist. Eventually, the company couldn't achieve targeted levels for sharing design expertise, components, and production processes, and overall product development costs rose.

Other companies have also been attempting to achieve mass customization with less than optimal results. Some of their experiences highlight the potential pitfalls companies can encounter in trying to make this leap.

- Nissan, Mitsubishi, and Mazda have run into many of the same problems that hurt Toyota. Nissan, for example, reportedly had 87 different varieties of steering wheels, most of which were great engineering feats. But customers did not want many of them and disliked having to choose from so many options.
- Amdahl, which built its business on a low-cost strategy but never made the move to continuous improvement, adopted a goal similar to Toyota's: deliver a custom-built mainframe in a week. However, Amdahl did not achieve its objective through flexible process capabilities, a dynamic network, or anything else resembling mass customization. It stocked inventory for every possible combination that customers could order, an approach that ended up saddling it with hundreds of millions of dollars in excess inventory.
- Dow Jones, through the *Wall Street Journal* and its other news-gathering resources, has a storehouse of information that it can customize and then deliver in a number of ways, including newswires, faxes, and on-line computer systems. Dow Jones, however, has not yet found the right formula for packaging services at a low price that would allow it to increase its share of the market. We suspect that two factors are responsible. Dow Jones seems to be trying to push a somewhat customized product out the door rather than first determining what customers truly need and how they want it delivered. The company also doesn't appear to have developed the organizational capabilities that would enable it to lower its costs enough to expand the still emerging market for customized information.

Despite the fact so many companies are struggling, scores of others are joining the quest. The appeal is understandable. Mass customization offers a solution to a basic dilemma that has plagued generations of executives.

Breaking New Ground

Until the widespread adoption of continuous improvement began about 15 years ago, either/or dichotomies dictated most managerial choices. A company could pursue a strategy of providing large volumes of standardized goods or services at a low cost, or it could decide to make customized or highly differentiated products in smaller volumes at a high cost. In other words, companies had to choose between being efficient mass producers and being innovative specialty businesses. Quality and low cost and customization and low cost were assumed to be trade-offs.

This old competitive dictum was grounded in the seemingly well-substantiated notion that the two strategies required very different ways of managing, and, therefore, two distinct organizational forms.

The *mechanistic* organization, so named because of the management emphasis on automating tasks and treating workers like machines, consists of a bureaucratic structure of functionally defined, highly compartmentalized jobs. Managers and industrial engineers study and define tasks, and workers execute them. Employees learn their jobs by following rigid rules under tight supervision.

In contrast, the *organic* organization, so named because of its fluid and ever-changing nature, is characterized by an adaptable structure of loosely defined jobs. These are typically held by highly skilled craftsmen. They learn through apprenticeships and experience, are governed by personal or professional standards, and are motivated by a desire to create a unique or breakthrough product.

The mechanistic organization, whether in a manufacturing or a service setting, gives managers the control and predictability required to achieve high levels of efficiency. The organic organization yields the craftsmanship needed to pursue a differentiation or niche strategy. Each of these organizational forms has innate limitations, however, which in the past have forced managers to choose one or the other. Almost all change is anathema to the mechanistic organization. And the artistry and informality at the heart of the organic organization defy efforts to regulate and control.

The development of the continuous-improvement and the mass-customization models show that companies *can* overcome the traditional trade-offs. In other words, companies can have it all.

Continuous improvement has enabled thousands of companies to realize lower costs than traditional mass producers and still achieve the distinctive quality of craft producers. But mass customization has enabled its adherents, which are as varied as Motorola, Bell Atlantic, the diversified insurer United Services Automobile Association (USAA), TWA Getaway Vacations, and Hallmark, to go a step further. These companies are achieving low costs, high quality, and the ability to make highly varied, often individually customized products.

Is Your Company Ready for Mass Customization?

Since achieving mass customization requires nothing less than a transformation of the business, managers must assess whether their companies must and whether in fact they can make the transformation.

Not all markets are appropriate for mass customization. Customers of commodity products like oil, gas, and wheat, for example, do not demand differentiation. In other markets, like public utilities and government services, regulation often bars customization. In some markets, the possible variations in services or products simply are of little value to customers. Also, variety in and of itself is not necessarily customization, and it can be dangerously expensive. Some consumer electronics retailers and supermarkets today are experiencing a backlash from customers confused by too broad a range of choices.

Continuous improvement will continue to be a very viable strategy for companies whose markets are relatively stable and predictable. But those companies whose markets are highly turbulent because of factors like changing customer needs, technological advances, and diminishing product life cycles are ripe for mass customization.

To have even a chance of successfully becoming a mass customizer, though, companies must first achieve high levels of quality and skills and low cost. For this reason, it seems impossible for mass producers to make the leap without first going through continuous improvement.

Westpac, the Australian financial services giant, is a case in point. It spent huge sums attempting to become a mass customizer by automating both the creation and delivery of its products. It wanted to install software building blocks that would allow it to create new financial products like mortgages and securities more quickly. Strategically, the move made sense. Deregulation had spawned a dizzying array of new products and services, and intensifying competition had caused significant downward pressure on prices.

Westpac tried to leapfrog continuous improvement by going from mass production directly to mass customization. The challenges of automating inflexible processes, building on ossified products, and trying to create a fluid network within a hierarchical organization—particularly at a time when the company was in poor financial condition due to intensifying competition in depressed markets—proved too difficult. Westpac has had to scale back significantly its ambitious dreams of becoming a tailored-product factory.

As we have stressed, even a company that has mastered continuous improvement must change radically the way it is run to become a successful mass customizer. A company must break apart the long-lasting, cross-functional teams and strong relationships built up for continuous improvement to form dynamic networks. It must change the focus of employee learning from incremental process improvement to generating ever-increasing capabilities. And leaders must replace a vision of "being the best" in an industry with an ideology of satisfying whatever customers want, when they want it.

The traditional mechanistic organization, aimed at achieving low-cost mass production, is segmented into very narrow compartments, often called functional or vertical silos, each of which performs an isolated task. Information is passed up, and decisions are handed down. Compensation of employees, who are viewed as mere cogs in the wheel, is generally based on standardized, narrowly defined job levels or categories.

In continuous-improvement organizations, the control system is much more, although never completely, horizontal. Increasingly, teams have not only responsibility for but also authority over a problem or task area. Such organizations are moving to much more generalized and overlapping job descriptions as well as to team-based compensation.

When mass customization is the objective, organizations structured around cross-functional teams can create horizontal silos just as isolated and ultimately damaging to the long-term health of the organization as vertical silos have been. When Toyota expanded dramatically its variety, for example, it found that tightly linked teams did not share easily across their boundaries to improve the general capabilities of the company. As a result, the costs of increasing variety rapidly outstripped any value it was creating for customers.

To achieve successful mass customization, managers need first to turn their processes into modules. Second, they need to create an architecture for linking them that will permit them to integrate rapidly in the best combination or sequence required to tailor products or services. The coordination of the overall dynamic network is often centralized, while each module retains operational authority for its particular process. Job descriptions become increasingly broad and may even disappear. And compensation for each module, whether it's a team or an individual, is based on the uniqueness and value of the contributions it makes toward producing the product.

Making Mass Customization Work

The key to coordinating the process modules is a linkage system with four key attributes.

1. *Instantaneous.* Processes must be able to be linked together as quickly as possible. First, the product or service each customer wants must be defined rapidly, preferably in collaboration with the customer. Mass customizers like Dell Computer, Hewlett-Packard, AT&T, and LSI Logic use special software that records customer desires and translates them into a design of the needed components. Then the design is quickly translated into a set of processes, which are integrated rapidly to create the product or service.

2. *Costless.* Beyond the initial investment required to create it, the linkage system must add as little as possible to the cost of making the product or service. Many service businesses have databases that make available all the information they know about their customers and their requirements to all the modules, so nothing new needs to be re-generated. USAA, for example, uses image technology that can scan and electronically store paperwork and a companywide database, so every representative who comes into contact with a customer knows everything about him or her.

3. *Seamless.* An IBM executive once commented, correctly, "You always ship your organization." What he meant was, if you have seams in your organization, you are going to have seams in your product, such as programs that do not work well together in a computer system. Since a dynamic network is essentially constructing a new, *instant team* to deal with every customer interaction, the occasions for "showing the seams" are many indeed. The recent adoption of case workers or case managers is one way service companies like USAA and IBM Credit Corporation avoid this. These people are responsible for the company's relationship with the customer and for coordinating the creation of the customized product or service. They ensure that no seams appear.

4. *Frictionless.* Companies that are still predominantly continuous improvers may have the most trouble attaining this attribute. The need to create instant teams for every customer in a dynamic network leaves no time for the kind of extensive team building that goes on in continuous-improvement organizations. The instant teams must be frictionless from the moment of their creation, so information and

communications technologies are mandatory for achieving this attribute. These technologies are necessary to find the right people, to define and create boundaries for their collective task, and to allow them to work together immediately without the benefit of ever having met.

USING TECHNOLOGY

In mechanistic organizations, the primary use of technology is to automate tasks, replacing human labor with mechanical or digital machines. People are sources of variation and are relatively costly, so mass producers often try to automate their companies as much as possible. This has the natural effect of reducing the numbers and skills of the work force.

In continuous-improvement companies, where workers are not only allowed but also encouraged to think about their jobs and how processes can be improved, technology is primarily used to augment workers' knowledge and skills. Measurement and analysis programs, computerized decision-support systems, videoconferencing, and even machine tools are aids, not people replacements.

In the dynamic networks of mass customizers, technology still automates tasks where that makes sense. Certainly, technology must augment people's knowledge and skills, but the elements of mass customization require that technology must also automate the links between modules and ensure that the people and the tools necessary to perform them are brought together instantly. Communication networks, shared databases that let everyone view the customer information simultaneously, computer-integrated manufacturing, workflow software, and tools like groupware (such as Lotus Notes) can automate the links so that a company can summon exactly the right resources to service a customer's unique desires and needs.

Many managers still view the promises of advanced technologies through the lens of mass production. But for mass customizers, the promise of technology is not the lights-out factory or the fully automated back office. It is used as a tool to tap more effectively all the diverse capabilities of employees to service customers.

While automating the links between modules is crucial, often some modules themselves can be automated by adopting, for example, a flexible manufacturing system that can choose instantly any product component within its wide envelope of variety. Motorola's Bravo pager factory in Boynton Beach, Florida, for example, can produce pagers—thanks to hardware and software modularity—in lot sizes as small as one within hours of an order arriving from a customer. The pager business is also a good example of how a mass customizer can automate links between modules. At Motorola, a sales rep and a customer design together, on a rep's laptop computer, the set of pagers (out of 29 million possible combinations) that exactly meets that customer's needs. Then the almost fully automated dynamic network takes over. The rep plugs the laptop into a phone and transmits one or more designs to the factory. Within minutes, a bar code is created with all the steps that a flexible manufacturing system needs to produce the pager.

As wonderful as these technological miracles sound, it is important to realize that technology is also potentially harmful. Mass customizers must periodically overhaul the linkages that they have adopted because as the market, the nature of their businesses, and the competition change, and as technology advances, any linkage system inevitably will become obsolete.

Another caveat: in this age when automated systems are handling daily millions of customer orders and inquiries placed via phones or computer systems, mass customizers must constantly be on their guard against eliminating their opportunities to learn what their customers like or dislike. Companies must always make it possible for their customers to "drop out" of the automated system so they can talk to a real person who is committed to helping them.

LEARNING FROM FAILURE

In the mechanistic organization, learning how to do something better is the prerogative of management and its collection of industrial engineers and supervisors. Workers only need to learn to do what is assigned to them; they don't have to think about it as well. The breakthrough of continuous improvement was the acknowledgment that workers' experience and know-how can help managers solve production problems and contribute toward tightening variances and reducing errors.

The differences between organizational learning in continuous-improvement and in mass-customization companies are most visible when you see how the two treat defects. Continuous-improvement organizations look at them as *process failures*, which the Japanese consider "treasures" because they provide the knowledge to fix problems and to ensure that failure never recurs.

In the dynamic networks of mass-customization organizations, defects are considered *capability failures:* the inability to satisfy the needs of some specific customer or market. They are still valuable treasures; but rather than sparking a spate of process-improvement activities, these defects call on the organization to renew itself by enhancing the flexibility within its processes, joining with another organization that has the needed capability, or even creating completely new process capabilities—whatever it takes to ensure that the customer is satisfied and, therefore, that capability failure doesn't happen again.

Capturing customer feedback on capability failures is crucial to sustaining any advantage that mass customization yields. A company that does this well is USAA, which targets its financial services and consumer goods to events in a customer's life, such as buying a house or car, getting married, or having a baby. Its information system allows sales reps to get customer feedback quickly on the phone and route it instantly to the appropriate department for analysis and action.

At Computer Products, Inc., a manufacturer of power supplies, marketing managers and engineers cold-call customers every day not to make a sale but to understand their problems and needs and to discuss product ideas. They then enter the information into a database that serves as an invaluable reference throughout the product-development cycle. Applied Digital Data Systems, a unit of AT&T's NCR subsidiary, uses a database system to store all its production information, including workers' comments and suggestions, and then regularly analyzes it to improve both its processes and products.

The capability to codesign and even coproduce products with customers provides mass customizers with the ability to capture valuable new knowledge. Motorola's and USAA's systems are good examples of this, as is Bally Engineered Structures Inc.'s. (See "Overcoming the Hurdles at Bally.") This is very different from what goes on in both mass-production and continuous-improvement organizations.

Typically, in those settings, there are almost no individual customer interactions that generate new knowledge.

Overcoming the Hurdles at Bally

In 1990, Tom Pietrocini, president of Bally Engineered Structures Inc. of Bally, Pennsylvania, decided that his company had to become a mass customizer to survive. He concluded that Bally had to change from a company that made specific products, like refrigerated rooms and walk-in coolers, to one that could make a growing range of products tailored to individual customers' needs but at the cost of standard mass-produced goods. Ideally, just what those products might be would be determined largely by how customers wanted Bally to use its hopefully ever-expanding set of capabilities to satisfy them.

It is too early to claim victory. With its markets still severely depressed, Bally is struggling. But its product range has widened to include even cleanrooms for the pharmaceutical industry, and today every product is tailored for each customer. Much more significant are the sweeping changes within the company that made the move to mass customization possible, such as the restructuring of process capabilities into modules that can be summoned in the numbers and combinations needed to create anything a particular customer seeks.

When Pietrocini joined Bally in 1983, it was a staid, high-cost mass producer that had been struggling to compete in a mature, cyclical industry soon to be wracked by price wars. It had positioned itself as a "quality" manufacturer, a strategy that had consigned it to the unenviable position of having to persuade customers to pay increasingly high premiums for its marginally better products.

Realizing this was untenable, Pietrocini started turning Bally into a lean, cost-efficient manufacturer that could grow by gaining market share. He hoped to achieve this by using the continuous-improvement approach to reduce significantly the number of defects and the time required to fill orders. Over several years, he broke down barriers between functions, created quality teams that were given wide latitude to make changes, and instilled within employees the belief that it was each person's responsibility not just to do a job but to figure out ways to do it better.

He developed an organization in which people talked to each other about production problems and enjoyed solving them together. They were driven by the vision of being the number one walk-in refrigerator company.

Bally was making considerable progress in enhancing its quality and bring-ing its costs in line with the rest of the industry when the last recession hit. That brought the already staggering industry to its knees. Before joining Bally, Pietrocini had worked in the auto-parts business, where he had twice ex-panded operations only to have to shrink them drastically during the OPEC-induced oil crises. After those painful experiences, he was loath to as-sume that Bally's market would fully rebound after the recession. He judged that continuous improvement alone would not be able to save the company and decided instead to remake it into an organization that thrives on the fickleness of customers and the turbulence of the markets.

Unlike many CEOs who have embarked on the road to mass customiz-ation only to stumble badly, Pietrocini committed his company with his eyes wide open. He understood that becoming a mass customizer would entail radical changes in organizational structure, systems, and culture. And during the last three years, he has succeeded at methodically transforming the $50 million company.

This transformation would not have been possible if Bally were still a mass producer. In those days, no one in manufacturing thought much about cus-tomers, let alone about their needs and wants. Aside from quality and cost levels that were out of kilter with the industry's, innovation—in terms of both products and processes—had stagnated at Bally. Although the continuous-improvement efforts gave Bally a fighting chance of making the leap to mass customization, the pillars of that approach were also obstacles that had to be removed. Even employees' perceptions of what Bally's business was got in the way of transformation.

As part of the continuous-improvement drive, Pietrocini had worked to convince employees that they were integral members of what was going to be "the best walk-in refrigerator company." That was fine when a good goal was continuing to do today what was done yesterday, only better. But this perception was an impediment to becoming a mass customizer. Workers had questioned, for instance, whether Bally should be making under-the-counter blast chillers, which process food instead of just storing it.

Now Pietrocini is trying to get employees to view the company in terms of its capabilities and values rather than as a maker of a concrete set of products. He preaches that things like efficiency, flexibility, and quality are the end rather than the means to achieving a rather limited purpose. He points out that cus-tomer demands and Bally's widening array of process capabilities will deter-mine what it will create.

Pietrocini also had to find ways to ensure that Bally's capabilities kept expanding. He encouraged employees to listen to and learn from every customer, and not just to depend on customer-reported defects or periodic customer-satisfaction surveys for feedback, as had been the tradition.

Bally's experience with one customer is a case in point. This customer decided to abandon Bally because his walk-in freezers' floors kept wearing out in as little as 18 months. Bally engineers discovered that the customer was cleaning the freezers with hot steam, something they were not designed to endure.

Bally would never have been able to advance the existing technology for making its products to the point where they could withstand that kind of punishment, but the company didn't simply write off the customer as unreasonable as it might have done in the past. Instead, a project team made up of people from various areas of the company worked on a solution part-time for two years. Ultimately, the team developed a completely new, patented technology that prevented moisture from entering crevices and destroying the floor. Bally not only won back this customer but also has used the technology for others.

Bally's structure also had to change radically to make the leap to mass customization. Before Pietrocini joined the company, Bally was producing modular panels that theoretically should have enabled it to customize individual orders. But the organization's structure was so rigid, it made doing so very difficult. Customers had to choose from a set number of standard product offerings in a catalog. Then, because of the bureaucratic way that the company processed orders, it took weeks just to get them to the shop floor. In addition, manufacturing processes were organized in a sequential order that left no room for modifications.

Bally has now not only broken up its tightly integrated set of processes but also has greatly expanded it. In the old days, when every order reached the factory floor, the foam panels, metal skins, corners, floors, ceilings, doors, and refrigeration units needed to cool the structure were built in largely that sequence. Since then, the number of customer options has soared from 12 to 10,000. And to create these options, Bally has greatly increased its number of process modules, so it can offer such features as welded construction and a much wider range of finishes and air- and electrical-control systems. Different modules are now called on to make each specific order, whether it is for a blast chiller, a clean room, or a freezer that can withstand steam cleaning.

The nervous system for Bally's new dynamic network is a sophisticated information-management system that Bally calls the computer-driven intelligence network (CDIN). A sales rep can custom-design each order in the customer's office on a laptop computer connected to the CDIN via a modem.

Once the design is completed, manufacturing software in the CDIN defines the precise combination of process modules required to make the product.

The network electronically connects everyone in the company, as well as independent sales reps, suppliers, and customers. The CDIN's databases contain nearly all the information Bally uses, including such things as leads, quotes, designs, purchase orders, and the skills and experiences of all Bally employees. This allows everyone to access the information they need to know without having to contend with functional boundaries. It also enables everyone to find quickly the people who have the skills they need for whatever issue that arises.

"We're a company of 400 people that behaves as if it were a company of 4 people, who talk to each other every day to decide how to best satisfy the customer of the moment," Pietrocini says.

CREATING A VISION

In addition to different attitudes about customer interactions, leaders of continuous-improvement companies and mass customizers foster very different approaches to the future. The former think they know what the organization needs to do to succeed in the future, whereas the latter believe that it's impossible to know and heresy to try because the future should be shaped by each successive customer order.

Leaders of continuous-improvement organizations provide a vision of not just what is to be done today but also what needs to be realized tomorrow, and this can work, provided that the market is relatively stable. Their vision is often expressed in terms of some competitive ideal of customer satisfaction. Allstate's "To be the best," Federal Express's "Never let the best get in the way of better," and Steelcase's "To provide the world's best office environment products, services, systems, and intelligence" are good examples. The common vision provides everyone in the company with the motivation, direction, and control necessary to continue improving all the time. Without a sustained vision, a company's attempts at process improvement can become lost in "program-of-the-month" fads or lip service to quality.

The highly turbulent marketplace of the mass customizer, with its ever-changing demand for innovation and tailored products and services, doesn't result in a clear, shared vision of that market. A standard product or market vision isn't just insufficient; it simply doesn't make sense. In a true mass-customization environment, no one knows exactly what the next customer will want, and, therefore, no one knows exactly what product the company will be creating next. No one knows what market-opportunity windows will open, and, therefore, no one can create a long-term vision of certain products to service those markets. But everyone does know that the next customer will want something and the next market opportunity is out there somewhere.

Many companies are articulating this scenario by using words like "anything," "anywhere," and "anytime." Peter Kann, chief executive of Dow Jones, describes his organization's strategic goal as providing "business and financial news and information however, wherever, and whenever customers want to receive it." Nissan's vision for the year 2000 is the "Five A's": any volume, anytime, anybody, anywhere, and anything. Motorola's pager group has a TV ad that asks, "How do you use your Motorola pager?" Various people answer with phrases like, "Anytime," "For anything," and "Anywhere I want."

No matter what they are called, such ideologies say two things about an organization: one, we don't know exactly what we'll have to provide to whom, and two, within our growing envelope of capabilities, we do know that we have or can acquire the capabilities to give customers what they want.

Leaders who can articulate such an ideology and create the dynamic network that can make it happen will succeed in moving their organizations far beyond continuous improvement to the new competitive arena of mass customization.

10
Managing by Wire

Stephan H. Haeckel and Richard L. Nolan

Flexibility and responsiveness now rule the marketplace. Rather than follow the make-and-sell strategy of industrial-age giants, today's successful companies focus on sensing and responding to rapidly changing customer needs. Information technology has driven much of this dramatic shift by vastly reducing the constraints imposed by time and space in acquiring, interpreting, and acting on information.

Responding to the competitive dynamic created by information technology, many large companies have drastically downsized, divested, and outsourced to reduce the costs and complexity of their operations. Yet simply reducing the size of a corporation is not the solution. As CEO Jack Welch has said, GE's goal is not to become smaller but to "get that small-company soul and small-company speed inside our big-company body." We believe that corporate size is worth saving. Market power, not bureaucratic clumsiness, can again become the dominant quality of a large corporation. But in order to survive in a sense-and-respond world, big companies must consider a strategy that we call *managing by wire*.

In aviation, *flying by wire* is a response to the changes introduced by jet-engine technology in the 1950s. It means using computer systems to augment a pilot's ability to assimilate and react to rapidly changing environmental information. Today heads-up displays (computer-generated pictures projected onto the pilot's helmet visor) present selected abstractions of a few crucial environmental factors, like oncoming aircraft and targets. Instrumentation and communication technologies aid in evaluating alternative responses. And when the pilot makes a decision, say to take evasive action by banking sharply to the left, it's the computer system that intercepts the pilot's command and translates it into the thousands of detailed orders that orchestrate the plane's behavior in real time.

When pilots fly by wire, they're flying informational representations of airplanes. In a similar way, managing by wire is the capacity to run a business by managing its informational representation. Manage-by-wire capability augments, instead of automating, a manager's function. Fly-by-wire technology—and by extension managing by wire—integrate pilot and plane into a single coherent system. The role and accountabilities of the pilot become an essential part of the design. Autopilot, or complete automation, is used only in calm, stable flying conditions. The system design allows for considerable flexibility in pilot behavior, including the ability to override the technology if, for instance, a sudden storm arises.

Like a plane at mach speeds, a company must be able to respond to threats in real time. In today's turbulent business environment, strategies have to be implemented in tactical timeframes. In response to this challenge, top-level managers need to view information technology in a new light. Rather than investing in isolated IT *systems*—such as e-mail, reservation systems, or inventory control systems—a company must invest in the IT *capabilities* that it will need to manage by wire.

The ideal manage-by-wire implementation uses an enterprise model to represent the operations of an entire business. Based on this model, expert systems, databases, software objects, and other technical components are integrated to do the equivalent of flying by wire. The executive crew then pilots the organization, using controls in the information cockpit of the business. Managers respond to the readouts appearing on the console, modifying the business plan based on changes in external conditions, monitoring the performance of delegated responsibilities, and sending directions to subsidiary units such as manufacturing and sales.

Of course, if the enterprise model represents the wrong reality—or is incomplete, out of date, or operating on bad data—the outcome could be catastrophic, like putting engines in reverse at 30,000 feet. Creating a robust model of a large business organization is an extremely challenging undertaking. But companies like Mrs. Fields Cookies, Brooklyn Union Gas, and a financial services organization that we will call Global Insurance have already demonstrated the feasibility of representing large portions of businesses in software. These companies manage by wire to varying degrees, from "hardwiring" automated processes at Mrs. Fields Cookies to a complete enterprise model that codifies business strategy at Global Insurance.

Many companies have spent decades automating pieces of their businesses, scattering networks and incompatible computer platforms throughout their organizations. But the empowered, decentralized teams of the information economy need a unified view of what's happening within an organization. Coherent behavior requires more than blockbuster applications and network connections; it must be governed by an enterprise model that codifies the corporation's intent and "how we do things around here." More important, a coherent model should include "how we *change* how we do things around here." Adding the institutional ability to adapt in a dynamic environment has become a survival imperative for most companies. And this ability will ultimately differentiate a manage-by-wire strategy from the static make-and-sell strategies of the past.

Hardwiring a Business

Over the last three decades, companies have used information technologies in increasingly sophisticated ways to run parts of a business. From the mainframe complexes of the 1960s to the client-server platforms of today, computers already help executives manage by automating business processes, from payroll to cash-dispensing. In fact, a company like Mrs. Fields can build an extensive representation of its business by automating procedures, that is, by codifying them in software.

In small companies, the model of "how we do things around here" often resides in the minds of a few people. Under these conditions, if senior executives are willing to sacrifice some flexibility and delegate the technical design to IT professionals, it's possible to represent enough of the business in software to manage by wire. For example, Mrs. Fields Cookies has captured a significant amount of its well-defined business in software. Its hardwired processes resemble the autopilot capability of a fly-by-wire system.

In 1978, when Debbi Fields opened her second cookie store in San Francisco (45 miles away from her first store in Palo Alto), she confronted the logistical problems of maintaining hands-on management at remote locations. She and her husband Randy, a skilled computer professional, had ambitious expansion plans that would prevent Debbi from personally overseeing each store. They needed a strategy that would let them know what was going on in hundreds of dispersed locations and at the same time ensure that local managers responded to daily challenges in the same way Debbi Fields would. In this case, Randy Fields had the technical expertise to implement in software the way Debbi Fields worked. He created the software at a reasonable cost and much more quickly than most traditional large-company IT groups could have.

Now with more than 800 stores, including franchises around the world, the central management of Mrs. Fields uses software to issue instructions and advice to store managers. Each morning, local managers project sales for the day and enter information into a personal computer: for example, day of the week, season, and local weather conditions. The software analyzes this data and responds with hourly instructions on what to do to meet the day's objectives: how many batches of different cookies to mix and bake; how to adjust the mixtures as the actual pattern of customer buying unfolds; when to offer free samples; how to schedule workers; and when to reorder chocolate chips.

There are a few fundamental principles that define Mrs. Fields's business concept: a thorough articulation of "how we do things around here"; a conviction that quality must be centrally controlled; and a dedication to knowledge sharing between central management and local store managers. As a matter of policy, the company integrates all of its information in one database and has one set of guidelines about how things are done the Mrs. Fields way. Because this vision is so clearly articulated, and because the company's business niche is relatively well-defined and stable, top management has, in effect, created an informational representation of Debbi Fields in each store.

Yet a manage-by-wire system that hardwires much of a business can turn out to be too rigid. For example, because its software was designed to describe the behavior of U.S. store managers, Mrs. Fields faced a number of challenges when it expanded into Europe and Asia, where different labor laws, languages, and supplier contracts had to be taken into account. In addition to the adjustments required to accommodate a wider range of local environments, falling profit margins forced the company to become more flexible in the way it applied information technology to running its daily business at remote locations.

Responding to these new conditions, Mrs. Fields Software (a separate business unit) developed a second generation of software, called the Retail Operations Intelligence system. ROI contained modules for inventory control, scheduling daily activities, interviewing and hiring, repair and maintenance, financial reporting, lease management, and e-mail. Senior management believes that ROI can be adapted to a variety of retail and service organizations. In fact, Mrs. Fields sold ROI to Burger King in 1992.

But at Mrs. Fields, top management relies on its IT division to translate business strategy into software. If senior executives want to change how the business runs, IT professionals must change the procedural software code. Because the cookie business doesn't change significantly from day to day—and employee turnover in a retail outlet like Mrs. Fields is high—it makes sense to run basic store operations as close to autopilot as possible. But most larger companies compete in more dynamic environments than Mrs. Fields, and, therefore, a corporate business model must do more than connect hardwired processes. It must also specify the roles and accountabilities of the people involved, incorporate the unplanned activity that can take up to 80% of a working day, and build in sufficient latitude for individual decision-making.

Institutionalizing Flexibility

Large organizations have become too complex for any individual, even the most brilliant executive, to keep complete models of the business in mind. Whether individually or collectively, managers of companies with hundreds of millions in revenue and tens of thousands of employees can't track everything that happens, much less coordinate millions of elements into a timely, coherent response. In fact, they never could, which is why functional hierarchies were originally created.

The old chain of command was designed for a relatively stable—and now increasingly rare—make-and-sell business. But many fast-growing sense-and-respond companies never adopted functional hierarchies in the first place. Instead, in the process of expanding, they have used IT-enabled networks as the tendons that hold the skeleton and muscles of the company together. Large companies, attempting to compete with agile niche players, are heading in the opposite direction of hardwiring operations. Rather than explicitly specifying "do it this way," many executives are empowering employees to "do it the best way you know how." However, without coordination, accountability, and shared objectives, this approach can often lead to paralysis rather than coherent company-wide behavior.

The need for flexibility drove the $1 billion utility, Brooklyn Union Gas of New York, to a radically different IT strategy. By the early 1980s, Brooklyn Union's 1971 Customer-Related Information System (CRIS) had become obsolete. Among other things, the Public Service Commission had begun requiring utilities to treat certain customers— for example, the elderly and disabled—in different ways. Top-level executives were also convinced that micromarketing increasingly customized service offerings was essential to Brooklyn Union's competitive survival. But the practices and policies of 1971 had become petrified in software procedures that were finally rendered obsolete by the dynamic environment that the company faced in the 1980s.

A $2 million initial attempt to upgrade CRIS failed. Finally, after spending more than three years on feasibility studies, design, and prototype systems, senior management agreed to let the IT department completely redo CRIS. The project began in the spring of 1987 and was completed by January 1990 at a cost of $48 million. In this case, the manage-by-wire implementation resulted not from a new business design by management but from the system being redesigned by a talented group of IT professionals.

The IT department chose to implement the new system using object-oriented programming. Objects are reusable software building blocks: sets of instructions that programmers can reassemble for a variety of different operations. CRIS now contains 650 such objects that create, in various combinations, 10,000 appropriate actions in 800 distinct business situations. These actions cover everything from meter reading and cash processing to collection, billing, credit, and field service orders. Brooklyn Union has now codified a substantial part of its customer-related business behavior in these software combinations. And because of the IT department's flexible, building-block approach to software, the system is much easier to modify than a hardwired one.

But at Brooklyn Union, as at Mrs. Fields, the IT department functions as the intermediary between customer-related management policy and its execution. The IT shop translates into software an understanding of management's business changes. It does this by defining the conditions that dictate legitimate combinations of software objects. These conditions may relate to business policy, legal requirements, or common-sense logic: for example, "You can't cut off service to an elderly customer before x months," or "You can't bill a customer if you haven't installed a meter."

Brooklyn Union exemplifies how computers can be used to create and manage building blocks of business activity that can then be combined and recombined into a variety of responses. However, senior executives are still disconnected from direct influence over the software that determines how their company handles customers. In fact, it is middle managers, rather than senior executives, who are managing by wire. And in the sense that the IT department acts as intermediary, Brooklyn Union has not moved beyond the practices of many large companies.

Not that top management at Brooklyn Union feels shortchanged. Its IT experts had the vision and ability to build in exceptional flexibility by using object-oriented software. As a result, new capabilities can now be created by extensively reusing existing software objects and adding only those required for specific additional functions. A proposal for a new engineering system, for instance, estimated that up to 30% of the software objects that were required to implement the system already existed in CRIS. More important, CRIS has delivered on top management's mandate against obsolescence, allowing Brooklyn Union to respond to market change and new opportunities in a timely way and at a reasonable cost.

Creating an Enterprise Model of a Business

Mrs. Fields's and Brooklyn Union's IT strategies demonstrate that manage-by-wire implementations vary from business to business, depending on size and complexity. A company's complexity is a function of how many information sources it needs, how many business elements it must coordinate, and the number and type of relationships that exist among those elements. We think of a company's *corporate IQ* as its institutional ability to deal with complexity, that is, its ability to capture, share, and extract meaning from marketplace signals. Corporate IQ directly translates into three IT infrastructure imperatives for connecting, sharing, and structuring information. (See Exhibit 10-1 "How 'Smart' Is a Company in a Complex World?")

Exhibit 10-1   How "Smart" Is a Company in a Complex World?

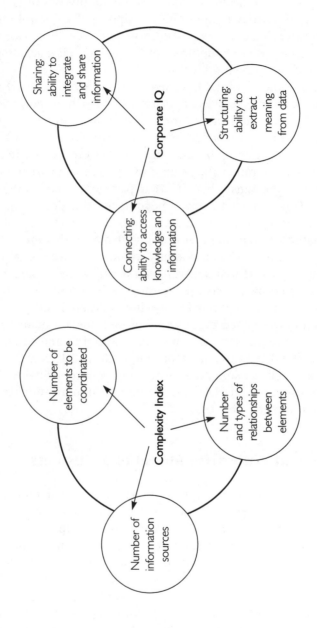

In most large companies, a low IQ results from change occurring so rapidly that keeping computer applications up to date is neither feasible nor affordable. Low IQs are particularly prevalent when processes have been automated over decades without any framework to integrate disparate applications and databases. At Mrs. Fields, where there are few information sources and clear and unchanging employee roles, ROI creates a high corporate IQ in an environment of comparatively low complexity.

Brooklyn Union's CRIS is less complete because it captures a smaller percentage of the total business. But this larger company operates in a much more complex industry. Brooklyn Union has a high capacity for sharing information and a comprehensive knowledge base in one important area: customers. Compared with many other large companies—with their disconnected information systems, competing computer platforms, and ill-defined business processes—Brooklyn Union Gas looks like a corporate genius.

But neither Mrs. Fields nor Brooklyn Union has a coherent model that fully maps key processes, how information is interpreted, and who is accountable for what. It is just such a model that can replace the IT department as the intermediary between management policy and execution. In fact, large companies need a coherent enterprise model to raise their corporate IQ.

An enterprise model is a high-level map of a business that guides the writing of computer code and the execution of nonautomated activities. Once procedures, data flows, and employee accountabilities are represented in computers by specific bit patterns and machine states, the map becomes the terrain; in other words, it becomes "real" in cyberspace, that computer-generated realm in which the informational representations of a cookie store or a utility's customer-related activities can be manipulated and modified. Companies can use an enterprise model to leverage a computer's memory and speed; to track and interrelate millions of events and relationships simultaneously; to allow selective sharing of information; and, finally, to initiate physical processes.

Of course, enterprise modeling tools have been available from software consultants and vendors for more than 25 years. Used primarily by information systems professionals to lay out procedures and data flows for certain business operations, the first generation of these tools were essentially high-level flow charts. Useful in highlighting procedural redundancies and omissions, they nonetheless have several major drawbacks that prevented their widespread adoption by management for designing business functions:

- They fail to incorporate the notions of commitment and human accountability in business processes, a particularly important omission because procedure without accountability often leads to bureaucracy.
- They don't deal with unstructured work and ad hoc processes.
- They take years to map into computer code, by which time the model is badly out of date.

Clearly, corporate managers, not IT professionals, should design a business. And business design extends beyond procedural design; it includes making strategic decisions about what market signals should be sensed, what data or analytical models should be used to interpret those signals, and how an appropriate response should be executed. To faithfully represent management's design, a robust enterprise model must consistently characterize any process at any scale, exhaustively account for the possible outcomes of every process, and unambiguously specify the roles and accountabilities of the employees involved in carrying them out.

A new generation of enterprise modeling tools that overcomes the drawbacks of traditional modeling tools is now emerging. Admittedly, creating a comprehensive information map is no simple task, but the benefits can be substantial, even for small business units. In a test at a large manufacturing company, one of these new enterprise modeling tools was used to map an engineering change process for electronic circuitry. Senior executives considered this process among the best in the organization. However, the new modeling tool not only revealed opportunities for procedural improvements, such as removing manufacturing bottlenecks, it also uncovered this startling fact: during the entire operation, not one person in the entire organization made a single commitment on volumes, cost, or delivery dates—only forecasts, estimates, and targets. If accountability isn't specified, business processes lack discipline and predictability, making them difficult to manage. A model that defines both procedures *and* accountability for outcomes can help managers of large companies do the job of managing.

The new enterprise modeling tools, for example, could make a substantial difference at Brooklyn Union Gas. CRIS uses data models to interpret signals from meter readings, field reports, and cash receipts. But the utility company has yet to develop an enterprise model that allows top managers to define and modify the policies that determine permissible combinations of its reusable software objects. An enterprise model would raise Brooklyn Union's corporate IQ by enhancing the structure of its customer information system. In effect, top-level managers would move into the information cockpit and gain the ability to modify directly how CRIS drives customer-related activities.

Designing the Intelligent Corporation

To be useful in today's dynamic business environment, an enterprise model must do more than represent a static version of "how we do things around here"; it must also include the capacity to adapt systematically and rapidly. Like the process of piloting a jet fighter, a true manage-by-wire system relies both on an accurate information model and on the organization's ability to learn.

The United States Air Force assesses a pilot's ability to learn with the OODA Loop, a model for the mental processes of a fighter pilot. OODA stands for:

- *Observation:* sensing environmental signals;
- *Orientation:* interpreting those signals;
- *Decision:* selecting from a repertoire of available responses;
- *Action:* executing the response selected.

Fighter pilots with faster OODA Loops tend to win dogfights, while those with slower ones get more parachute practice. Note that the loop is iterative: a continuous cycle in which an action leads to the observation of the results of that action that in turn requires a new orientation, decision, and action. This iterative sequence constitutes a *learning loop*. It contains the four functions essential to any adaptive organism: sensing, interpreting, deciding, and acting. By analogy, an enterprise model for a business that incorporates learning is one that systematically creates and links learning loops. (See Exhibit 10-2 "Learning Loops.")

Recent work on organizational learning focuses on the way that people in a company learn. But what about *institutional* learning? How much do companies know when the people go home at night? Many companies, with the aid of software, would know how to process payrolls. Some would know how to dispense cash and others how to replenish stocks. But one could hardly call that learning.

Exhibit 10-2   Learning Loops

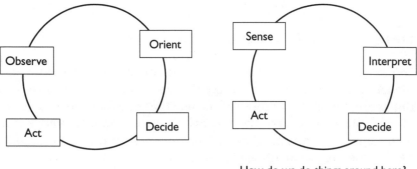

We define institutional learning as the process by which information models change, be they data models, forecasting models, or procedural models. Therefore, a good enterprise model should include a design for systematically changing these kinds of models, based on signals received from the environment. That means an adaptive organization avoids running learning loops repeatedly over static information models.

An example of an institutional learning loop at work is the system that Wal-Mart and its apparel suppliers use to replenish stocks in Wal-Mart stores. For instance, every evening, Wal-Mart transmits five million characters of data about the day's sales to Wrangler, a supplier of blue jeans. The two companies share both the data and a model that interprets the meaning of the data. They also share software applications that act on that interpretation to send specific quantities of specific sizes and colors of jeans to specific stores from specific warehouses. The result is a learning loop that lowers logistics and inventory costs and leads to fewer stock outs. And every time the data model is changed to reflect a new fashion season or pricing pattern, both Wal-Mart and Wrangler learn and adapt.

Using technology to integrate how an organization interprets "what's going on out there" with a codification of "how we do things around here" creates an intelligent corporation. The company we call Global Insurance (the real company is disguised) is one of the best examples of managing by wire at this level of sophistication.

A large financial services organization, Global was driven to a fundamental reconceptualization of how it does business because of competition from niche players in the 1980s. The $78 billion company was facing extinction. New policies took two years from conception to consumer, and operational costs were 15% higher than those of smaller competitors, who were luring customers away with innovative offerings. Furthermore, the insurance industry was changing so rapidly that senior executives had little confidence that any specific strategy would keep the company afloat for more than a year or two.

In the late 1980s, rather than investing in a specific business strategy, top-level managers decided to spend $110 million on an IT infrastructure that would allow them to implement *any* strategy quickly. Senior executives started the project with the development of an enterprise model for the company's two largest business lines: casualty and life insurance. The model linked product development, underwriting, sales, and other functions in a coherent informational representation of "how we do things around here." Information specialists created the model based on senior managers' specifications of the information they wanted to track (observation); the data models needed to interpret the information (orientation); the analytic and decision support provided for underwriters, actuaries, and managers in the field (decision); and, finally, how these decisions should be executed via their on-line transaction systems (action).

Using combinations of more than 1,000 software objects, the company created the transactions, activities, and data that would, when properly linked, define any present or future offering in its life- or casualty-insurance business lines. "Enroll client," "send premium notice," and "establish risk limit" are examples of these software building blocks.

In addition, data models were developed to interpret the market research, transaction history, demographic, and economic information that Global collected from the field, external databases, and internal operations. Data models are explicit renderings of the way an application program, or a collection of these programs, views the world. When these models are used to create databases, they institutionalize specific ways of interpreting raw data. Elaborate data models are worth fortunes to banks, airlines, food manufacturers, and large retailers like Wal-Mart, because they help these companies reorient themselves continually.

At Global, a decision-support system used the patterns its data models revealed to trigger exception reports or approval requests that then appeared on managers' terminals. For instance, a manager whose product was losing ground to a new competitive offering would have the option of modifying the existing policy or creating an entirely new one. This process was codified by expert systems that contain legal, logical, and business constraints: for example, "we will not underwrite an aggregate risk for a single client that is more than x times the client's net worth."

Through this manage-by-wire capability, decision-to-action times have been reduced by 400% to 700%, enabling Global to meet the competition of small niche players. And certain types of decisions can now be carried out in real time. One example: agents, who have their own laptop computers with easy access to Global's electronic network, can customize a policy in a client's living room. They can tailor policies based on a client's specific situation, such as annual income, ages of dependents, or lifestyle preferences.

There was, of course, substantial technological risk associated with Global's project. Still riskier was counting on the ability of managers to specify adequately the hundreds of procedures and dozens of management policies necessary to ensure that Global's responses were consistent with its business goals. The CEO also worried that his information systems team, lacking sufficient business experience, might misinterpret these specifications when they translated business rules into the language of data models and expert systems. It was only through extensive prototyping that senior executives acquired the confidence to transfer processes gradually to the manage-by-wire system.

Along the way, Global has experienced setbacks. Vendor technology was late and slow. Some managers, who implicitly relied on bureaucratic procedures to buffer them from direct accountability for policy changes, resisted the extensive retraining that was designed to put them in the pilot's seat. In fact, many managers didn't make the shift successfully. Some key executives retired or left the company, taking with them crucial knowledge that the rest of the institution hadn't learned because it had never been codified.

But after more than a year's delay and a budgetary overrun, Global has implemented almost all it set out to do technologically. Because its enterprise model wasn't developed with the new generation of modeling tools, changes to the model must still be made by the IT shop. Still, the senior executives who run the life and casualty businesses are managing by wire a large portion of their operations. With a few additional changes, managers will be able to modify underwriting policy themselves through the IT system and have these changes reflected immediately in the policies written by agents.

Setting Guidelines for Managing by Wire

Given the right enterprise model and a technology-enabled capacity to learn, a large company's size can again become a decisive competitive advantage. But to many managers, Global Insurance's successful manage-by-wire strategy will seem unattainable. For one thing, the technical expertise needed to implement such an integrated system may not exist within the company. The ability to change reality by modifying an informational representation of it is possible only with an underlying technological infrastructure that has a high corporate IQ. Indeed, managing by wire requires the long-term commitment of both senior executives *and* a world-class IT group.

Flying a modern jet airplane is a sophisticated operation. The current generation of fly-by-wire systems requires more than 20 million lines of computer code. Yet if an aviation information model can successfully capture this level of complexity, an enterprise model can do the same for the managers of rapidly changing business units. In fact, adopting a manage-by-wire strategy is nothing less than a change in the nature of strategy itself, from a *plan* to produce specific offerings for specific markets to a *structure* for sensing and responding to change faster than the competition.

Faced with an unpredictable business environment, top managers at Global Insurance were forced to fuse their business and IT strategies. It's imperative that today's senior executives make IT policy an integral part of corporate strategy and intent. Technological knowledge must join the financial and operational know-how of a policy-making manager; otherwise, crucial business decisions will implicitly be delegated to the IT department.

Managers can follow a few guidelines to help them implement a manage-by-wire system:

Top managers must assess a company's corporate IQ in terms of connecting, sharing, and structuring information. There are three critical attributes of a company's IT infrastructure that determine its corporate IQ. *Connecting* means the degree to which the IT platform links information sources, media, locations, and users. Since the 1970s, computer networks have sprung up in multiple places for multiple purposes. As a result, many companies today are crisscrossed by dozens of independent networks that are incompatible technically and thus actually inhibit, rather than promote, information sharing. Mere connectivity doesn't necessarily increase productivity *or* institutional learning. Management must not only determine what the signals should be but also ensure that these signals are understood and shared by the right people and teams.

Sharing makes possible coordinated effort and, therefore, the benefits associated with teamwork, integration, and extended scope. Getting everyone on the same page in a large business requires an institutional capability to share data, interpretations of that data, and specifications of core processes. The added value that this integration can yield underlines a subtle but important distinction: the actual implementation of a breakthrough application, such as an automated airline reservation system like American Airlines's SABRE, may ultimately be less important than *how* that application is implemented. A stand-alone application is less likely to deliver sustainable competitive advantage than one implemented on an integrated technology platform designed for extensive information sharing. Anyone who receives multiple premium notices on the same day from the same insurance company for different policies is on the receiving end of an unintegrated IT platform.

Structuring holds the most potential for the strategic exploitation of information in the 1990s and beyond. Structure is created by information about information, for instance, how data is classified, organized, related, and used. Tables of contents, indices, and "see also" references are familiar hard-copy examples. The data models of Global Insurance and Wal-Mart structure information by filtering the data that bombards these companies every day.

When information from previously unrelated sources is structured in a meaningful way, human beings become capable of thinking thoughts that were previously unthinkable. Computers that use their speed and memory to reveal patterns in raw data augment the extraordinary capacity of humans to recognize and assign meaning to patterns. For example, through spectral analysis and mathematical equations that model what scientists call the red shift, a computer can process light signals from a remote galaxy to calculate the distance and size of its parts. The results can be displayed in a three-dimensional picture and then rotated. Presentation in this manner allows scientists to "see" a distant galaxy from the back or side and even to discover, as they did recently, a huge void passing through it.

An enterprise model should be expressed in business language, not IT terminology. Management should select and use one business design language and insist on its use throughout the organization. In many companies, a variety of first- and second-generation enterprise modeling tools have already been used to capture key processes in different functions or operating units. But in order to create a unified understanding of "how we do things around here" (and, if it makes strategic sense, to facilitate future integration of presently autonomous organizational units), a common business language is required.

Senior executives must determine the highest level at which coherent institutional behavior adds value. Managers must decide which business units, if well coordinated, could together create more value than the sum of their individual parts. In many respects, this is the strategic task facing managers in a sense-and-respond world. There's no one answer to this crucial issue. Many different approaches have been tried, even in an information-intensive industry like publishing. McGraw-Hill's strategy, for example, is to treat their information systems and certain editorial content as assets to be shared among multiple units. Dun and Bradstreet, on the other hand, views its information and technology as assets to be separated into individual units. In other words, McGraw-Hill shares assets at the enterprise level and Dun and Bradstreet at the business-line level.

Once a company embarks on a manage-by-wire strategy, senior executives must carefully plan the pace of its implementation. Just as information technology has fueled a new competitive dynamic for businesses, the advent of jet-engine technology in the 1950s profoundly affected aviation. By increasing the speed of fighter planes, the jet engine made it impossible for pilots to fly planes manually. But flying by wire didn't happen overnight. In the mid-1950s, no pilot would have felt safe with a sudden and comprehensive introduction of software between the cockpit controls and the physical airplane, even if the technology had existed at the time. In fact, only the latest generation of commercial aircraft truly fly by wire.

Similarly, few executives will feel confident enough to commit their company to managing by wire in one massive effort. How fast and how far they're willing to go will depend on how effectively the software currently mediates management decisions; how much confidence managers have in their IT staff; and how much money and time it will ultimately take to implement the process.

When a target level for coherent institutional behavior has been defined, common information and technology assets can be leveraged to create economies of scope. But realistically, most companies will model smaller business domains first, such as Brooklyn Union's customer information system. They will then link these domains to cover larger parts of the business, as Global did.

No corporation has implemented a fully integrated manage-by-wire system yet. But a growing number of companies like Brooklyn Union Gas and Global Insurance are showing that large and complicated business operations can be captured in an information technology structure and used to govern business behavior. These companies have already significantly improved their response times and substantially reduced the costs of developing new products and services.

The imminent arrival of a new generation of enterprise modeling tools makes a manage-by-wire strategy plausible. But it will be management's skill in codifying a competitive information model that will determine its success.

Executive Summaries

The Emerging Theory of Manufacturing

Peter F. Drucker

We cannot yet build the factory of 1999. But we can already see how it will have to be built and managed. Four concepts—Statistical Quality Control (SQC), the new manufacturing accounting, the "flotilla" organization of the manufacturing process, and systems design—are transforming how we think about manufacturing. Each is being developed separately, by different people with different agendas. But together they provide the foundation for the new theory of manufacturing that manufacturing people know we badly need.

Traditional approaches to manufacturing—scientific management, the assembly line, cost accounting—see the parts, not the whole. The factory is simply a collection of machines, isolated from the rest of the business process (distribution, service, and the like). In the manufacturing theory now emerging, the factory is little more than a wide place in the stream of producing value. "Manufacturing" decisions and "business" decisions are fundamentally one and the same.

Each concept achieves its own form of integration. SQC aligns information with accountability to put control of work in the hands of the operators. The new accounting theory redefines fundamentals (like costs and benefits) to reflect the realities of automated production. The flotilla, or module, organization of manufacturing gives us low cost *and* variety by combining standardization and flexibility. Systems design forces us to recognize that producing does not stop when goods leave the factory dock.

Together these concepts tackle the conflicts that have been most troublesome for twentieth-century mass-production factories: conflicts between people and machines, time and money, standardization and flexibility, functions and systems.

Marketing in an Age of Diversity

Regis McKenna

The old days of mass production and mass marketing are over, says Regis McKenna. Today technology has combined with a fragmented culture to create a dizzying array of products, services, and markets. It is a world of variety and options, niches and small batches, increased competition and changing company structure.

From his experience helping some of the country's most innovative high-tech ventures develop and market products, Mr. McKenna offers these observations on today's fractured marketplace:

- The decline of branding, the rise of "other." Branding no longer guarantees a loyal customer; increasingly, "other"—small nonbranded products and companies—is winning market share.
- The false security of market share. Managers should wake up every morning uncertain about the marketplace—because it is invariably changing.
- Niche marketing: selling big by selling small. Because niche markets cannot be identified easily in their infancy, managers must keep one foot in the technology to know its potential and one foot in the market to see the opportunity.
- The integrated product. Today the product isn't just the thing itself—it's also service, financial reports, type of technology, even the personality of the CEO.
- The customer as customizer. With technologically driven products, the customer can practically invent the market for a company.
- The evolution of distribution. Today the product is an experience the customer learns to trust—which is why everything from perfume to software samples come inside magazines.
- Goliath plus David. Established giants and small startups need each other—the future is in these productive relationships that create products tailored to customers' demands.

Managing in an Age of Modularity

Carliss Y. Baldwin and Kim B. Clark

Modularity is a familiar principle in the computer industry. Different companies can independently design and produce components, such as disk drives or operating software, and those modules will fit together into a complex and smoothly functioning product because the module makers obey a given set of design rules.

Modularity in manufacturing is already common in many companies. But now a number of them are beginning to extend the approach into the design of their products and services. Modularity in design should tremendously boost the rate of innovation in many industries as it did in the computer industry.

As businesses as diverse as auto manufacturing and financial services move toward modular designs, the authors say, competitive dynamics will change enormously. No longer will assemblers control the final product: suppliers of key modules will gain leverage and even take on responsibility for design rules. Companies will compete either by specifying the dominant design rules (as Microsoft does) or by producing excellent modules (as disk drive maker Quantum does).

Leaders in a modular industry will control less, so they will have to watch the competitive environment closely for opportunities to link up with other module makers. They will also need to know more: engineering details that seemed trivial at the corporate level may now play a large part in strategic decisions. Leaders will also become knowledge managers internally because they will need to coordinate the efforts of development groups in order to keep them focused on the modular strategies the company is pursuing.

Do You Want to Keep Your Customers Forever?

B. Joseph Pine II, Don Peppers, and Martha Rogers

Customers, whether consumers or businesses, do not want more choices. They want exactly what they want—when, where, and how they want it—and technology now makes it possible for companies to give it to them. But few companies are exploiting that potential. Most managers continue to view the world through the twin lenses of mass marketing and mass production. To handle increasingly turbulent and fragmented markets, they try to churn out a greater variety of goods and services and to tailor their messages to ever finer market segments. But they end up bombarding their customers with too many choices.

A company that aspires to give customers exactly what they want must use technology to become two things: a mass customizer that efficiently provides individually customized goods and services, and a one-to-one marketer that elicits information from each customer. The process of acquiring those skills will bind producer and consumer together in what the authors call a learning relationship—an ongoing collaboration to meet the customer's needs over time that will continually strengthen their bond.

In learning relationships, individual customers teach the company about their preferences and needs. The more they teach the company, the better it gets at providing exactly what they want and the harder it becomes for a competitor to entice them away. A company that can cultivate learning relationships with its customers should be able to retain their business forever, provided that it continues to supply high-quality customized products or services at competitive prices and does not miss the next technology wave.

Is Your Company Ready for One-to-One Marketing?

Don Peppers, Martha Rogers, and Bob Dorf

One-to-one marketing, also known as relationship marketing, promises to increase the value of your customer base by establishing a learning relationship with each customer. The customer tells you of some need, and you customize your product or service to meet it. Every interaction and modification improves your ability to fit your product to the particular customer. Eventually, even if a competitor offers the same type of service, your customer won't be able to enjoy the same level of convenience without taking the time to teach your competitor the lessons your company has already learned.

Although the theory behind one-to-one marketing is simple, implementation is complex. Too many companies have jumped on the one-to-one bandwagon without proper preparation—mistakenly understanding it as an excuse to badger customers with excessive telemarketing and direct mail campaigns.

The authors offer practical advice for implementing a one-to-one marketing program correctly. They describe four key steps: identifying your customers, differentiating among them, interacting with them, and customizing your product or service to meet each customer's needs. And they provide activities and exercises, to be administered to employees and customers, that will help you identify your company's readiness to launch a one-to-one initiative.

Although some managers dismiss the possibility of one-to-one marketing as an unattainable goal, even a modest program can produce substantial benefits. This tool kit will help you determine what type of program your company can implement now, what you need to do to position your company for a large-scale initiative, and how to set priorities.

Breaking Compromises, Breakaway Growth

George Stalk, Jr., David K. Pecaut, and Benjamin Burnett

Many companies today are searching for growth. But how and where should they look? Breaking compromises can be a powerful organizing principle. Even in the most mature businesses, compromise breakers have emerged from the pack to achieve breakaway growth—far outpacing the rest of their industry. Examples include Chrysler Corporation, Contadina, CarMax, and the Charles Schwab Corporation.

Compromises are concessions customers are forced to make. Unlike trade-offs, which are the legitimate choices customers make between different product or service offerings, compromises are imposed. For instance, in choosing a hotel, a customer can *trade off* luxury for economy. But the entire hotel industry makes customers *compromise* by not permitting early check-in. Trade-offs are very visible, but most compromises are hidden.

Compromises mean it's the industry's way or no way. Often, customers assume the industry must be right; they accept compromises as the way the business works. That is why traditional market research rarely uncovers compromise-breaking opportunities.

The authors propose a number of alternative approaches to finding the compromises hidden in any business. One approach is to look for the compensatory behaviors customers engage in because using the product or service as intended would not fully meet their needs. Other approaches include paying attention to performance anomalies and looking for diseconomies in the industry's value chain. If managers think like customers, the authors say, they will be able to find and exploit compromises for faster growth and improved profitability.

The Four Faces of Mass Customization

James H. Gilmore and B. Joseph Pine II

Virtually all executives today recognize the need to provide outstanding service to customers. Focusing on the customer, however, is both an imperative and a potential curse. In their desire to become customer driven, many companies have resorted to inventing new programs and procedures to meet every customer's request. But as customers and their needs grow increasingly diverse, such an approach has become a surefire way to add unnecessary cost and complexity to operations.

Companies around the world have embraced mass customization in an attempt to avoid those pitfalls. Readily available information technology and flexible work processes permit them to customize goods or services for individual customers in high volumes at low cost. But many managers have discovered that mass customization itself can produce unnecessary cost and complexity. They are realizing that they did not examine thoroughly enough what kind of customization their customers would value before they plunged ahead. That is understandable. Until now, no framework has existed to help managers determine the type of customization they should pursue.

James Gilmore and Joseph Pine provide managers with just such a framework. They have identified four distinct approaches to customization. When designing or redesigning a product, process, or business unit, managers should examine each approach for possible insights into how to serve their customers best. In some cases, a single approach will dominate the design. More often, however, managers will need a mix of some or all of the four approaches to serve their own particular set of customers.

Versioning: The Smart Way to Sell Information

Carl Shapiro and Hal R. Varian

Many producers of information goods assume that their products are exempt from the economic laws that govern more tangible goods. But that's just not so. Information goods are subject to the same market and competitive forces that govern the fate of any product. And their success, too, hinges on traditional product-management skills: gaining a clear understanding of customer needs, achieving genuine differentiation, and developing and executing an astute positioning and pricing strategy.

What makes information goods tricky is their "dangerous economics." Producing the first copy of an information product is often very expensive, but producing subsequent copies is very cheap. In other words, the fixed costs are high and the marginal costs are low. Because competition tends to drive prices to the level of marginal costs, information goods can easily turn into low-priced commodities, making it impossible for companies to recoup their up-front investments and eventually bringing about their demise.

The best way to escape that fate, the authors say, is to create different versions of the same core of information by tailoring it to the needs of different customers. Such a "versioning" strategy can enable a company to distinguish its products from the competition and protect its prices from collapse.

The authors draw on a wide range of examples to illustrate how companies use different versioning strategies to appeal to customers with different needs. The power of versioning is that it enables managers to apply tried-and-true product-management techniques in a way that takes into account both the unusual economics of information production and the endless malleability of digital data.

Making Mass Customization Work

B. Joseph Pine II, Bart Victor, and Andrew C. Boynton

Scores of companies, including Toyota, Amdahl, Dow Jones, and Motorola, have been trying to become mass customizers: businesses that produce individually customized goods or services at the cost of standardized, mass-produced goods. The appeal is understandable. Generations of executives have believed that they could not customize *and* have low costs and high quality. But mass customization is a way to have it all.

Many aspiring mass customizers, however, have stumbled. Toyota, for example, had to retreat from its goals of offering customers a wide range of options and delivering a made-to-order car within three days.

Toyota—and other companies—did not realize that mass customization is not simply an advanced stage of continuous improvement. True, companies must first achieve high quality and low costs, and they must develop a highly skilled, flexible work force capable of handling a large degree of complexity. But successful mass customization calls for more: total transformation of the organization.

Mass customization entails breaking up the tightly integrated networks that form the backbone of the continuous-improvement organization and creating a loosely linked collection of autonomous modules. Each module performs a different task and is perpetually reconfigured in response to customer demands. Automation typically is the key to linking these modules so that they can come together quickly and efficiently.

Leaders of mass-customization organizations never know exactly what customers will ask for next. All they can do is strive to be more prepared to meet the next request. To that end, mass customizers are forever changing and expanding their range of capabilities.

Managing by Wire

Stephan H. Haeckel and Richard L. Nolan

Rather than follow the make-and-sell strategies of industrial-age giants, today's successful companies focus on sensing and responding to rapidly changing customer needs. Information technology has driven much of this dramatic shift by vastly reducing the constraints imposed by time and space in acquiring, interpreting, and acting on information. In order to survive in this sense-and-respond world, big companies need to consider a strategy that Stephan Haeckel and Richard Nolan call *managing by wire.*

In aviation, *flying by wire* means using computer systems to augment a pilot's ability to assimilate and react to rapidly changing environmental information. When pilots fly by wire, they're flying informational representations of airplanes. In a similar way, managing by wire is the capacity to run a business by managing its informational representation.

Rather than investing in isolated IT *systems,* a company must invest in the IT *capabilities* that it will need to manage by wire. Indeed, coherent corporate behavior needs more than blockbuster applications and network connections; it must be governed by a coherent information model that codifies a corporation's intent and "how we do things around here." More important, a coherent model must include "how we change how we do things around here."

Companies like Mrs. Fields Cookies, Brooklyn Union Gas, and a financial services organization that the authors call Global Insurance are managing by wire to varying degrees, from "hardwiring" automated processes at Mrs. Fields to a complete enterprise model that codifies business strategy at Global Insurance.

About the Contributors

Carliss Y. Baldwin is the William L. White Professor of Business Administration at Harvard Business School, and coauthor, with Kim Clark, of the book *Design Rules: The Power of Modularity*. Prior to her position at Harvard, she taught at MIT's Sloan School of Management. She is currently involved in a multiyear project with Kim Clark to study the process of design and its impact on the structure of the computer industry. In addition to research and teaching, Professor Baldwin serves as a Director of Country Curtains, Inc.

Andrew C. Boynton is a Professor of Management and the Director of the Executive M.B.A. Program at the International Institute for Management Development (IMD) in Lausanne, Switzerland. He has previously held faculty positions at IMD, the Darden School at the University of Virginia, and the Kenan-Flagler Business School at the University of North Carolina at Chapel Hill. Boynton has written numerous articles on strategy organization transformation and the competitive use of information for *Harvard Business Review, Sloan Management Review,* and *California Management Review.* He is the coauthor of *Invented Here: Maximizing Your Organization's Internal Growth and Profitability.*

Benjamin Burnett is Vice President of The Boston Consulting Group (BCG) in Chicago. He is a member of BCG's Consumer Goods and Retail and Financial Services Practice Groups. Mr. Burnett has worked with a number of packaged goods and retail clients in developing corporate strategies, building brands, driving trade execution, conducting competitive analyses, and driving innovation and growth. In particular, he has helped clients identify sources of growth through deeper consumer understanding.

Kim B. Clark is the Dean of the Faculty and the George F. Baker Professor of Administration at Harvard Business School. His most recent book, *Design Rules: The Power of Modularity,* is coauthored with Carliss Baldwin and focuses on modularity of design and the integration of technology and competition in industry evolution, with a particular focus on the computer industry. His earlier research has covered the areas of technology, productivity, product development, and operations strategy. He has published numerous articles in journals such as *Harvard Business Review, California Management Review, Management Science,* and *Administrative Science Quarterly,* and is the author or coauthor of seven books.

Bob Dorf is President of the Peppers and Rogers Group. He has led the firm's consulting practice and managed the firm for five years, spearheading the one to one needs assessment implementation planning processes. He is coauthor, with Don Peppers and Martha Rogers, of *The One to One Fieldbook.* Dorf was Founder and CEO of Dorf & Stanton Communications, a communications and public relations firm, until 1989. He is a member of the Public Relations Society's honor society and a former editor of both *Marketing and Media Decisions* magazine and WINS 1010 Newsradio in New York.

Peter F. Drucker is a writer, teacher, and consultant whose 31 books have been published in more than 20 languages. He is the Founder of the Peter F. Drucker Foundation for Nonprofit Management, and has counseled numerous governments, public service institutions, and major corporations. *Peter Drucker on the Profession of Management* collects some of his best *Harvard Business Review* articles in one volume.

James H. Gilmore is Cofounder of Strategic Horizons LLP of Aurora, Ohio, a thinking studio dedicated to helping businesses conceive and design new ways of adding value to their economic offerings. He is a member of the Creative Education Foundation and the Creative Thinking Association of America, and frequently helps managers learn specific innovative thinking techniques to enhance their personal creativity. He is the inventor of the Sacrifice-Mapping technique and co-author, with partner Joe Pine, of *The Experience Economy: Work Is Theatre & Every Business a Stage.*

Stephan H. Haeckel is Director of Strategic Studies at IBM's Advanced Business Institute and Chairman of the Marketing Science Institute. As IBM's Director of Advanced Market Development, he led the formulation of IBM's corporate strategies for alternate marketing channels and for entering the commercial systems integration business. At the ABI, Haeckel teaches and advises executives on the design, implementation, and leadership of adaptive organizations. Haeckel has served on the Advisory Council of the Federal National Mortgage Association and the panel of judges for the McKinsey Awards, and represents IBM on the Marketing Council of the American Management Association. He published *Adaptive Enterprise: Creating and Leading Sense-and-Respond Organizations* in 1999.

Regis McKenna is Chairman of The McKenna Group, an international strategy consulting firm specializing in the application of information and telecommunications technologies to e-business strategies. He is on the board of The Economic Strategies Institute and the Competitiveness Council. McKenna is also Chairman of the Board of the Santa Clara University Center for Science, Technology and Society and was a founding board member of Smart Valley. He is a Trustee at Santa Clara University, on the advisory board of the Haas School of Business at the University of California at Berkeley and the Carnegie Mellon University Computer Sciences School, and President of the board of trustees for The New Children's Shelter of Santa Clara County. He has helped numerous entrepreneurial start-ups during their formation years, including America Online, Apple, Compaq, Electronic Arts, Genentech, Intel, Linear Technology, National Semiconductor, Silicon Graphics, and 3COM. His most recent book is *Real Time: Preparing for the Age of the Never Satisfied Customer.*

Richard L. Nolan is the William Barclay Harding Professor of Management of Technology at Harvard Business School. Nolan returned to the faculty of Harvard Business School in 1991, after serving as Chairman of Nolan, Norton & Co. since 1977. He is currently studying business transformation, the process of creatively destroying industrial economy management principles and evolving workable management principles for the information economy. He is coauthor, with David

Croson, of *Creative Destruction: A Six-Stage Process for Transforming the Organization* and, with Thomas Davenport, Donna L. Stoddard, and Sirkka Jarvenpaa, *Reengineering the Organization*. His latest book, edited with Stephen P. Bradley, is *Sense and Respond: Capturing Value in the Network Era*.

David K. Pecaut is Coleader of The Boston Consulting Group's global electronic commerce practice, in which he helps BCG clients respond to opportunities and threats in the fast-moving world of e-commerce. Pecaut acts as both strategist and business architect, creating new business models for the e-commerce space and bringing together prospective partners to build them. His work focuses on issues of competitive strategy, new business development, branding, pricing, fulfillment, and implementation in electronic commerce. Pecaut also manages the Canadian operations of BCG.

Don Peppers is Cofounder and Partner of the Peppers and Rogers Group. He is coauthor, with Martha Rogers, of *The One to One Future* and *Enterprise One to One*, seminal works in the rapidly growing fields of customer relationship management (CRM) and interactive marketing. The duo's newest book, *The One to One Fieldbook*, coauthored with Bob Dorf, President of the Peppers and Rogers Group, was released in January 1999. Peppers is also the author of *Life's a Pitch—Then You Buy*, based on his own career as a new business rainmaker for advertising agencies including Chiat/Day and Lintas: USA. He and Martha Rogers were named 1998 Direct Marketers of the Year by Direct Marketing Days in New York.

B. Joseph Pine II is Cofounder of Strategic Horizons LLP of Aurora, Ohio, a thinking studio helping executives see the world differently. He is a Faculty Leader in the Penn State Executive Education Program, a member of the executive education faculty at the UCLA Anderson Graduate School of Management, and an adjunct faculty member with the IBM Advanced Business Institute. He is also the author of *Mass Customization: The New Frontier in Business Competition* and coauthor, with partner Jim Gilmore, of *The Experience Economy: Work Is Theatre & Every Business a Stage*.

Martha Rogers is Cofounder and Partner of the Peppers and Rogers Group. Her 1993 book with Don Peppers, *The One to One Future*, created the theories behind customer relationship management (CRM). Rogers is also coauthor with Peppers of *Enterprise One to One* and coauthor of *The One to One Fieldbook,* with Peppers and Bob Dorf. A Professor at Duke University Fuqua School of Business and a member of the Dean's Advisory Council at Indiana University, Rogers was named the 1997 International Sales and Marketing Executives' International Educator of the Year. Her work with Peppers has appeared in *Harvard Business Review, Forbes ASAP, Wired, Marketing Tools, BrandWeek, New Media Age* (U.K.), and *Diamond New Media* (Japan). Peppers and Rogers were named 1998 Direct Marketers of the Year by Direct Marketing Days in New York.

Carl Shapiro is the Transamerica Professor of Business Strategy at the Haas School of Business at the University of California at Berkeley. He is also Director of the Institute of Business and Economic Research and Professor of Economics at UC Berkeley. He has been Editor of the *Journal of Economic Perspectives* and a Fellow at the Center for Advanced Study in the Behavioral Sciences. Shapiro served as Deputy Assistant Attorney General for Economics in the Antitrust Division of the U.S. Department of Justice during 1995–1996. He also founded the Tilden Group, an economic consulting company, and is now a Senior Consultant at Charles River Associates. His book with Hal R. Varian, *Information Rules: A Strategic Guide to the Network Economy,* was published in 1998.

George Stalk, Jr., is a Senior Vice President of The Boston Consulting Group and focuses his professional practice on international and time-based competition. He speaks regularly to business and industry associations on time-based competition and other topics. Based in Toronto, he has served as a consultant to a variety of leading manufacturing, retailing, and technology- and consumer-oriented companies. Mr. Stalk is the coauthor of the critically acclaimed *Competing Against Time* and *Kaisha: The Japanese Corporation,* and has published articles in numerous business publications.

Hal R. Varian is the Dean of the School of Information Management and Systems at the University of California at Berkeley. He also holds joint appointments in the Haas School of Business and the Department of Economics and occupies the Class of 1944 University Professorship. Varian has published numerous papers in economic theory, econometrics, industrial organization, public finance, and the economics of information technology.

Bart Victor is the Cal Turner Professor and Director of the Cal Turner Program in Moral Leadership at the Owen Graduate School of Management at Vanderbilt University. Prior to joining the Owen School, he was a Professor of Management on the faculties of the International Institute for Management Development (IMD) in Lausanne, Switzerland, and the Kenan-Flagler Business School, University of North Carolina. Victor's research interests include strategic innovation and growth, moral leadership, and high performance leadership. He consults with firms on leadership and strategy creation in the United States and Europe.

Index